The Survival of a Mathematician

From Tenure-Track to Emeritus

The Survival of a Mathematician

From Tenure-Track to Emeritus

Steven G. Krantz

AMERICAN MATHEMATICAL SOCIETY

2000 *Mathematics Subject Classification.* Primary 00–01; Secondary 00A99, 00A30, 00A05, 00A06.

For additional information and updates on this book, visit
www.ams.org/bookpages/mbk-60

Library of Congress Cataloging-in-Publication Data

Krantz, Steven G. (Steven George), 1951–
    The survival of a mathematician : from tenure-track to emeritus / Steven G. Krantz.
      p. cm.
    Includes bibliographical references.
    ISBN 978-0-8218-4629-2 (alk. paper)
    1. Mathematics—Vocational guidance. 2. Mathematics—Study and teaching (Higher)
3. Universities and colleges—Professional staff. 4. Career development. I. Title.

QA10.5.K734   2008
510.23—dc22   2008036232

To all the advice I never got and had to figure out for myself.

Table of Contents

Preface

A few years ago I wrote and published a book with the American Mathematical Society called *A Mathematician's Survival Guide* [KRA3]. This volume was intended to help the student learn how to become a mathematician. Feedback from readers has indicated that this book has found an appreciative audience, and it has been successful in mentoring young mathematicians. The book helps the newly minted Ph.D. to find his/her place in the mathematical firmament and to learn how to get along in the profession. My motivation for writing that book was a commonly held belief or observation that mathematics is traditionally a sink-or-swim vocation; there is nobody to tell you what you are supposed to (or are expected to) do in your new position, how you are to learn the ropes, and how you are to advance and realize your potential. There is some truth to this claim, but two comforting facts are that this differs little from the challenge facing most people as they embark on their careers and that there is guidance to be found for those who seek it.

Certainly the transition from the intensity and often solitary activity of getting a thesis written to the sociopolitical structure of an academic or industrial job can be a shock to the system. Nothing in your formal education prepares you for the many nuances and loopholes of your new work environment. You will have many new choices as to what sort of working environment you should select; if you are fortunate, you will find one that suits your interests and proclivities. This could be a first-rate academic/research environment, or it could be a primarily teaching environment, or it could be in a genome lab, or with a computer firm, or with a branch of the federal government. In every instance you will face similar questions: What am I supposed to do (on a daily basis, and also in the long run)? How do I function effectively and successfully in this new setting? What are my goals? What is expected of me? To whom am I answerable?

On the face of it, the Ph.D. is preparation for a research career. The fresh Ph.D. should be chomping at the bit to prove theorems and write papers. But it is a hard fact that most American Ph.D. mathematicians write very few papers. According to recent statistics from the American Mathematical Society, of those authors who publish anything at all in their careers:

- About 43% publish only one paper

- About 15% publish only two papers

- About 8% publish only three papers

- About 75% publish five or fewer papers

Many authors publish just a paper based on the thesis and nothing more. Why is this? Is the cutting of the (academic) umbilical cord so traumatic that most people just fall off the wagon? Or are the reasons more complicated? Do people just get wrapped up in other duties, or other career pursuits, and decide after a while that "publish or perish" is not part of their credo? Are they perhaps in jobs in which publishing and doing research is not really the thing that is rewarded?

And what about teaching? If you are working for the National Security Agency (as, for instance, three of my Ph.D. students now are), then you certainly will not be teaching classes, grading papers, or giving grades. But you *will* have to give seminars. You *will* have to mentor others. You *will* have to provide guidance to younger staff members. How does one learn these skills?

And, no matter where you work or what you do, you will no doubt work as part of a team. You will have to function in meetings and on conference calls and in interactions with your supervisors and your subordinates.

If you are in an academic job, then your role(s) in life is carefully delineated and described in your institution's *Tenure Document*: teaching, research, and service are the three branches of an academic's professional activity. He/she is judged on each of these, and in different ways. For example, if you manage to prove the Riemann hypothesis, then it doesn't matter whether you spend your time at staff meetings rolling your eyes and humming *The Battle Hymn of the Republic*. If you are a world-class teacher, then you will probably be granted some slack in your research program. If you are a terrific departmental citizen, seen as a person who holds the ship of state together, then you will perhaps not have to put in quite so much time on the other two portions of your profile.

The bottom line is that there is an awful lot about this profession that you are going to have to figure out for yourself. This book is intended to help you through the process. One of the main messages here is to *talk to people.* Find a senior faculty member who is willing to let his/her hair down and tell you some things about how life works in your department, or your organization, or your company. Bond with others who are your peers and who can share experiences with you. Become friendly with the staff, with the Chair, with the key players in your group or department. I can assure you that—if you are in an academic department—a good deal of the decision of whether to tenure you is based on raw quality, but another good part of it is based on collegiality and whether you will fit in. Is this someone that we want to have knocking about in this building for the next forty years or not? Is this someone whom we would look forward to seeing each day? These are intangibles, not written in any guidebook or *Tenure Document.* But they are facts of life.

The purpose of the present volume is to give you some hints as to how to make your way in the academic world, or more generally in the corporate world or professional world of mathematics. I cannot claim to be expert in every nuance and corner of the profession; but I have had more experience than most. I can certainly help you to avoid most of the pitfalls.

I should perhaps stress that I know quite a lot about the life of a mathematician in the United States. I know very little about that life in other countries. I do know that there can be considerable differences—in culture, in style, and in emphases. I must leave it to another scholar to write a book about the mathematical life in Italy or Sri Lanka.

By the same token, almost all of my professional experience has been of an academic nature. I have done some consulting, and I have collaborated with nonmathematicians. So my communication skills are moderately well developed. But I have never worked for Microsoft, or at the Social Security Administration, or in a genome laboratory. And I probably never will. I know some of the key features of non-academic jobs, and I intend to share them here. But it is a foregone conclusion that the focus of this book will be largely on an academic career.

It is a pleasure to thank Gerald B. Folland and James S. Walker for a careful reading of an early draft of this book, and for contributing many useful and incisive comments. Robert Burckel, in his unbeatable style, studied every word that I wrote and corrected them accurately and mercifully. David Collins painstakingly taught me the chapter and verse of the *Chicago Manual of Style* (or CMS), and many other truths about language and accuracy as well. Ed Dunne, as always, was an encouraging and dynamic editor. He read several drafts of the book and contributed decisively to its form and structure. Ed also engaged five excellent reviewers who contributed incisive criticisms and suggestions that have certainly made the book tighter and more on point.

Mathematics is a highly varied, rich, and rewarding life. Welcome to it. I hope that you spend a very pleasant and productive thirty or forty years making your way through the profession, and that you find many rewards and comforts. May this book be your touchstone as you get started.

— Steven G. Krantz
St. Louis, Missouri

Part I

Simple Steps for Little Feet

Chapter 1

The Meaning of Life

Machiavelli's teaching would hardly have stood the test of Parliamentary government, for public discussion demands at least the profession of good faith.

> Lord Acton (British literary figure)

A life which does not go into action is a failure.

> Arnold J. Toynbee (historian)

I think one of the greatest joys I have now in my career and in my profession is to be playing at an age where I can appreciate it more than I used to . . . It's a whole different lens you look through the older you get.

> Andre Agassi (tennis player)

The profession had a profound saddening effect on my life.

> Armand Assante (actor)

In England, the profession of the law is that which seems to hold out the strongest attraction to talent, from the circumstance, that in it ability, coupled with exertion, even though unaided by patronage, cannot fail of obtaining reward.

> Charles Babbage (mathematician, scientist)

The ABC of our profession is to avoid these large abstract terms in order to try to discover behind them the only concrete realities, which are human beings.

> Marc Bloch (historian)

1.1   What Am I Supposed to Be Doing Here?

And well might you ask. When I landed my first job—an Assistant Professorship at UCLA—I may as well have been placed as first trombone in the Milwaukee Symphony. I had no clue of who I was or where I was supposed to be or what I was supposed to do. Well, that is not quite true. I knew that I was a math professor and that I was supposed to teach classes and to prove theorems. But I had no detailed knowledge of what that really entailed.

Certainly the first thing you should do when you show up in your new department, assuming that you are in an academic job, is to go to the Chair's office and introduce yourself to people. This includes all the secretaries and the staff and, of course, the Chair or Head himself/herself. Be prepared to sit for a while and pass the time of day with the Chair—be sure that you have enough time to get acquainted! Discuss your duties, your goals, and your frame of mind as you join this new department. Ask the Chair whom you should meet, who will be the key people in your life.

You will also want to find out who is the Vice-Chair for Undergraduate Studies and the Vice-Chair for Graduate Studies and introduce yourself to those people. You may not have meaningful relationships with these folks for a while yet. But they are, or will be, significant players in your life. You want to know who they are, and you want them on your team. Spend a little time studying the entire composition of the department and its place in the university infrastructure. There may be a Coordinator of Lower Division Teaching, a Supervisor of Undergraduate Advising, a Graduate Student Mentor, and any number of other people whom you never dreamed of before.[1] They are all a part of your world now, and you would do well to get to know them all—at least to the extent of being able to say hello to them when you meet them in the hall.

An immediate need and responsibility for you is to find out who are the key people in your subject (i.e., research) area. Knock on their doors. Introduce yourself. Find out when the seminar meets and become an active and participating member. That means that you should volunteer to give talks, you should attend all the meetings, you should participate enthusiastically and meaningfully. If the analysts are all in the habit of going out for a beer

---

[1]If you are in a small department, with just a handful of faculty, then the structure of the department will be much simpler. Certainly there will not be so many officers. You may find, in such a context, that you are inheriting responsibilities much faster than you expected. This matter will be discussed later in the book.

on Friday afternoons, and if you are an analyst, then you had best join in. If there is an intramural soccer team, then you probably ought to throw your hat in the ring.

Things will be different at different types of institutions. At a smaller college, the individual departments are somewhat small, and there is a good deal of interaction among the different departments. Mathematicians routinely have lunch with faculty from engineering or French or history, and they have friends in many different disciplines. Life at a comprehensive university will be somewhat like this as well. So if your new job is at a place like this, then you will start meeting a variety of faculty, with a variety of different backgrounds, early on. There may not be a seminar in your subject area—or even an active research group. But, if you are lucky, there may be a larger, research-oriented institution not far away (within a hundred-mile radius, let's say) where you could go for seminars and some mathematical talk.

The fact is that, at a teaching college or a comprehensive university, your focus is going to be a bit different. Now you are going to want to get to know everyone, because you will be interacting with everyone on a regular basis. Such departments are generally run rather democratically, and you want to make an effort to fit in from the get-go. It is quite common for a small department to have a multihour faculty meeting two or three times per week. This is where departmental business is dispatched and many decisions are made. It takes the place of a myriad of departmental officers and committees. So your job now is to figure out the system and become a part of it.

When I was at UCLA, all the movers and shakers in the math department participated in a monthly poker game. It was by invitation only, and I was never invited. But this was where many of the most important departmental decisions were made. It was the proverbial "smoke-filled room" where deals were made and broken. If you were part of it, then you were a "made man." Otherwise not.

This is life. What appears on the surface of things, what is written in the university catalogue, what is written in the *Tenure Document*, is only the tip of the iceberg when it comes to understanding how the place really works and how the power structure really functions. It is essential that you develop a good, working relationship with a senior mentor—someone who can give you regular reality checks on how things are going in the department, and particularly *how you are doing* in the department. How can you find such a person and get to know him/her? More will be said about this matter as the book develops. Certainly attending seminars, going to teas, attending social events, and cultivating mathematical conversations are obvious ways to start. Some departments or organizations will actually *assign* you a senior mentor the day that you walk in the door. In my own department we found that this didn't work very well because it was a bit artificial. Most times you will have to identify and develop a relationship with such a person yourself.

1.2   Getting to Know You

I have already said that you must get to know people. If you land at your new job and just hide in your office, then your future will not be bright. You may be chuckling in your beard, but in fact it's all too easy in an academic job to just teach your calculus classes and then go home. That is a sure recipe for failure.

You really want to become a fixture around the department. You want the staff to like you and to think of you as someone that they can depend on. You want the senior faculty to look forward to seeing you each day, to look forward to hearing about your (mathematical) results. When a senior faculty member goes to another school to give a colloquium, or goes to a conference, he/she should be saying to his/her friends, "We've got this terrific new young guy/gal in our department. He/she is a real plus to our program, and a gifted young mathematician. We were lucky to hire him/her."

If you are at an institution where teaching is the major emphasis, then perhaps you will establish your bona fides in the department in a slightly different fashion. Become a knockout calculus teacher. Come up with innovative ways to get your students involved in the subject matter. Hold special problem sessions. Create special software for your course. The main point is to find a way to make your presence known. You want everyone to know who you are and what you have to contribute.

I don't mean to downplay your potential relationship with other junior faculty or junior staff. These are really your comrades-in-arms, and you want to get to know them too. Certainly don't think of yourselves as competitors for some mutually exclusive holy grail. It's not as though if Bob gets tenure then the slot is gone so you will be denied tenure. *Usually* tenure is a zero-one game that you play against yourself. That is, if you make the grade, you get tenure, and if you don't make the grade, you don't. It happens only very occasionally (contrary to what you may see in a Hollywood movie) that a department will be told that for budgetary reasons it can only tenure one person this year—even though it has three good candidates who are ready for tenure.

These days there are some very useful and active organizations that help young mathematicians and, more generally, young scholars get oriented in their new professional lives. One of these is the Young Mathematicians' Network (YMN).[2] Located at `http://concerns.youngmath.net/`, this is an organization founded by a group of young mathematicians who wanted to create a resource for people looking for jobs, people trying to get settled in a new department, people trying to get tenure. Going to the Website, you will see that YMN sponsors conferences, hosts Websites and discussion groups, and mentors activities around the country. Most of the founders of this enterprise now have tenure in some good departments around the country, and the torch has been passed to a new generation. But the activity continues, and it is certainly valuable and worthwhile. In fact it has spawned the book [BEC], and this is a fine resource for the beginning mathematician.

Another excellent touchstone for the beginning mathematician, or more generally the beginning scholar, is Project NExT (New Experiences in Teaching), sponsored by the ExxonMobil Foundation and a number of other companies and organizations. Project NExT is overseen and administered by the Mathematical Association of America. This is a loose-knit organization of junior faculty across the country who want to share common interests and concerns. They are mentored by a broad cross-section of senior mathematicians who make themselves available for consultation or for just chewing the rag. The Project NExT people have their main meeting each year at the Summer MathFest (sponsored by the MAA); then they reconvene at a smaller event at the January AMS/MAA meetings. They also organize other special events. Project NExT endeavors to inform its members about publishing, about tenure, about teaching, and about getting along in a math department.[3] It has done a lot of good for a lot of people, and I encourage you to get involved—the Website is `http://archives.math.utk.edu/projnext/`.

---

[2]This group has also become known as "Concerns of Young Mathematicians" (CYM).

[3]The young mathematician's home department is required to be a part of Project NExT. In particular, it is the home department that pays for travel to the NExT meetings.

1.3   Getting to Know Your Teaching

Teaching is exciting and rewarding and can also be fun. Interacting with bright young people is certainly one of the finer things in life. Explaining important ideas to a receptive audience is fulfilling, and is also important for bringing a new generation of young adults up to speed in our discipline. You are fortunate to be part of a vocation that puts you front and center in this process. Make the most of it.

What does this mean? First of all, you will get a whole lot more out of your teaching—and everyone else will too—if you are reasonably good at it. The ability to teach well is *not* something you are just born with—like the ability to hear with perfect pitch. It is a cultivated skill, and one that you should start working on right away—see Section 2.1. Some of the traits of a good teacher are simply matters of tending to business: You prepare your lectures carefully, you write a good syllabus, you choose an appropriate and readable text. Other traits are special and personal and will require hard work.

You will probably have had some experience as a teaching assistant, or TA, and that is an activity that resembles teaching. But *really teaching*—being in charge of a class, writing the exams, assigning the grades, handling the problem situations—is a rather more sophisticated activity.

I may humbly suggest that you consult the book [KRA1], which will give you the full story on almost every aspect of teaching, and more particularly of teaching mathematics. God is in the details, and you will find that the enterprise of teaching is certainly a whole that is greater than the sum of its parts. *Preparation* is a big part of being an effective teacher. You want to convey the immediate and powerful impression that you are a professional who is on top of the material and who knows how to communicate it. Many of your other shortcomings will be forgiven, or at least overlooked, if it is clear that you are a pro who is doing his/her best to do a top-notch job. You want to be courteous, kind, and fair. I have always gotten along well with my classes and garnered reasonably good teaching evaluations,[4] but in recent years I have done even better than usual because students warm up to the fact that I am so easygoing. I think this means that when they come to me with a problem—a forgotten assignment, or an overslept exam, or a

---

[4]There are a few exceptions, such as the teaching evaluation that said that I should not be allowed to teach any biped in any state west of the Mississippi.

plane ticket that conflicts with the final, or some other completely irrational, unjustifiable quagmire of a situation—I always say, "OK, we can probably handle that. Let's sit down and work something out." I have found over the years that such an attitude requires no more effort, and is no more of a strain, than chewing the student out, or trying to create more trouble for everyone.

There are particular skills to writing a good exam, to grading the exams fairly, to determining course grades, and so forth. It requires some genuine insight to assess a class, determine the students' level and preparation, and then pitch the lessons so that the students will understand them and benefit from them. Again, these matters are addressed in some detail in [KRA1]. Good teaching is a skill that you will hone over a period of years, just like a good golf game or a good attack on the cello. Talking to colleagues, both your senior mentors and your peer junior faculty, is an extremely valuable exercise. It is always useful to bounce your ideas off of others with a similar set of experiences. Sometimes you can do a thought experiment with your friends and thereby avoid a cataclysm in the classroom.

Whether you hang your hat in a research department or a teaching department or (like my own) a department that is a mixture of both, you will do well to have a positive teaching reputation. You will thereby have the respect and admiration of your students and colleagues, and a definite plus in your portfolio. It is unlikely that you will get tenure just on the basis of your teaching alone, but teaching will certainly play a key role in the decision. Indeed, in most math departments today, if you are a top-notch researcher but a distinctly lousy teacher, then you will almost certainly not get tenure.

1.4 Getting to Know the Other Aspects of Your Life

I shall say repeatedly in this book that the three big vectors in an academic life are

- teaching

- research

- service

Of course one of the main messages of the book is that there really is a lot more to it than that simple list. But those three are milestones, and I shall say a great deal in the ensuing pages about them.

Service is in some sense the easy part of your life, because you don't even have to think about it. It will be thrust upon you. That is to say, you will live your ordinary life in the math department, and you will be assigned certain committee or task force duties. And you will do them—presumably responsibly and effectively. For most of us, that is the extent of service. You can be asked to serve on university-wide committees, and you should do so with your usual aplomb and professionalism. You might be tapped to be Vice-Chair for Undergraduate Studies or Vice-Chair for Graduate Studies or even Chair (i.e., Chairperson or Head of the department). To these you should not give a knee-jerk "yes" answer, because any of these is a big commitment. On the one hand, you feel an obligation to serve your colleagues and your institution. On the other hand you have a life to live. You may have a spouse or significant other, and a family, and perhaps a church or other religious affiliation. You need to balance all the components of your life. Subsequent sections of this book will discuss the various aspects of these different types of service and what they entail.

Perhaps the most difficult—and also the most rewarding—of the three components indicated above is research. It is difficult because most likely nobody has told you how to build your own research career, how to forge a path in the research world, how to establish a research identity. To get to the nitty gritty, *how do you find problems that are worth working on and how do you solve them and how do you write them up and how do you get them published?* This is the essential question to answer if you want to establish a scholarly reputation and get tenure in a good department. All of Chapter 4 is devoted to different aspects of the research life and how to cope with them.

This book tries to paint your life as a tapestry with many warps and wefts. You need to get along with many different types of people and you need to master many different kinds of tasks. And do so gracefully and with skill. If you can make this happen, then you will lead a rewarding and productive life, and you can write your own version of this book for the next generation.

1.5 Collegiality

In the 1950s, 1960s, and even most of the 1970s, math departments were extraordinarily friendly places. Salaries were low, duties were many, but the attitude was "we're all in this together". The lovely book [DAVH] captures the spirit of the camaraderie of the time.

It was very common in those days for there to be a colloquium each week, followed by a fairly large and high-spirited colloquium dinner, followed by a party at someone's house. When I was an Assistant Professor at UCLA we had all these features, often followed by a swim in Richard Arens's pool.

In fact I can recall many a time when, after lunch, one of the guys (and this time I really do mean a *guy*) would phone home and say, "Hello, dear. Joe Schlomokin from Purdue is in town. He's giving a talk. Nobody else is giving the party, so I thought we could do it. Could you run to the store and pick up some stuff? Also he needs a place to flop and I told him he could sleep on our sofa. We'll be going to dinner, and we'll show up for the party at 8:00 p.m." Miraculously, the spouse would reply with suitable enthusiasm, and the festivities would begin in due course.

Times have changed. Today most spouses work. Many spouses work as academics, and often in the same department as the other spouse. So there are a lot of shared responsibilities: child rearing, cooking, soccer game coaching, and on and on. This means that attendance at colloquium dinners is much thinner. This also means that there is nobody to phone up and tell to go out and pick up stuff for an impromptu party. And so forth. There are very few colloquium parties anymore—except for very distinguished or special guests.

In fact two-career couples often make imaginative accommodations to the issues raised in the preceding paragraph. For one thing, planning ahead for parties is much more common. Having cross-disciplinary parties—to honor someone from English at the same time as someone from math—is a new and often pleasing development. Obviously *both* members of the couple must pitch in for all aspects of the party, or for any other entertaining (dinners, outings, picnics, etc.) that is done. And certainly accommodations must be made for the kids, the pets, or perhaps an aging parent who has become part of the household.

Collegiality takes on new meaning, and has new practical significance, when a two-career couple with kids is trying to play the game. It is still of utmost importance to be collegial, to be friends with your colleagues, to have a proactive and trusting relationship with the people in your workplace. This certainly will involve activities other than actually working (much of the work that a mathematician does is, after all, solitary). But you will actually have to work at being collegial.

Also the discipline has become more competitive. In the old days nobody was paid very well, and almost everyone with a body temperature above 93° had an NSF research grant. Today salaries are all over the map—and everyone knows it—and NSF grants are about as hard to get as vintage Elvis Presley records. Often the department colloquium has disintegrated into a number of competing seminars.

I don't mean to paint a bleak picture. Math departments can still be friendly places—fun to work in and intellectually stimulating. But they are different from what they were in years past.

In the late 1970s at UCLA there was a very special logic seminar called the Cabal Seminar. One might wonder about the provenance of this unusual name. Certainly it suggests something dark and mysterious for the cognoscenti. It turns out that the seminar was named in honor of the organizers' favorite real estate agent. Whenever they used her services to help a new mathematician relocate, she would give them a kickback from her commission. And they used the money to run the seminar.

The pleasures derived from this largesse were quite evident. On Fridays, when the rest of us were at tea eating Ritz crackers and drinking tepid tea, the logicians would be sitting off in the corner quaffing chilled wine and eating Camembert and pâté de foie gras. And they were able to bring in a number of classy speakers for their mathematical activities.

This is just the reality of life. There was nothing unfriendly about what the logicians were doing. But those who have enjoy and those who don't have don't.

It is important to do what you can to contribute to the collegiality of your Department. Go to lunch with colleagues. Participate in Ping-Pong games or intramural sports. Go out with friends for a beer after work. Get together on weekends for barbecues or picnics. Give as many parties as you (and your partner) feel comfortable giving. Working with people whom you like and trust, and with whom you feel comfortable, is a commodity that you just cannot buy. It can really smooth things out in your professional life.

1.6 What Else Is There to Life?

Well, more than you ever imagined. I have been a Professor now for thirty-four years, and my mother still thinks that all I do is teach. When I tell her that my teaching load is typically two courses per term, she wonders what I do with the rest of my time. I am tempted to say that I coach the football team.

An academic mathematician is *not* a high school teacher. While teaching is a *very important* part of what you do, it is by no means the only thing. Measured by the number of hours you will put into it, teaching is well less than half of what you do.

The rest of what you do is **(i)** research, **(ii)** exposition, **(iii)** departmental administrative activities, **(iv)** University administrative activities, **(v)** service to the profession. In item **(iii)** I am including, of course, serving on committees and running programs, but also activities that relate to teaching, such as undergraduate advising. Which is important if you care about math majors and the program overall. In item **(v)** I include editing of journals, refereeing, service on national committees, attending national meetings such as the January joint meeting of the AMS/MAA. In item **(iv)** I include anything that the Dean or the Provost or even your Chair may ask you to do. In item **(ii)** I include survey articles, book reviews, textbook writing, and any of the other myriad writing activities that one may take on in this line of work.

The life of an academic mathematician is rich and complex. You should read this entire book to get a palpable feeling for all its many dimensions. You must try to keep all the different components in perspective, and make some decisions about how to apportion your time. If you give all your time to teaching, then you will not be able to develop your research profile. If you give all your time to research, then your other activities will suffer. You want to be passionate about all the different aspects of your life, and you also want to give each its due.

Chapter 2

Your Duties

Apathy can be overcome by enthusiasm, and enthusiasm can only be aroused by two things: first, an ideal, which takes the imagination by storm, and second, a definite intelligible plan for carrying that ideal into practice.

Arnold J. Toynbee (historian)

What a beautiful art, but what a wretched profession.

Georges Bizet (composer)

So I'm in quite the wrong profession obviously.

Dirk Bogarde (actor)

Your morals and general character are strictly inquired into; it is therefore expected that you will improve every leisure moment in the acquirement of knowledge of your profession and you will recollect that a good moral character is essential to your high standing in the Navy.

Franklin Buchanan (Naval officer)

In our film profession you may have Gable's looks, Tracy's art, Marlene's legs or Liz's violet eyes, but they don't mean a thing without that swinging thing called courage.

Frank Capra (movie director)

2.1 How to Teach

Whole books and many articles have been written on the art and practice of good teaching (see, for instance, [KRA1], [CAS], [DAV], [GKM], [MOO], [REZ], [RIS], [ROSG], [THU], [TBJ], [ZUC], [STSA]). It is a gentle and delicate yoga, one that you will (if you are smart and dedicated) hone and perfect for your entire professional life. Being a good teacher is like being a good parent, or a good spouse, or a good friend. It is not something you are going to learn and perfect just by reading a book. Certainly the book can give you some useful pointers and help you to avoid pitfalls. But, in the end, you are going to recognize that this is quite a personal activity and that you will need to develop your own values, your own goals, and your own methods.

The Carnegie Foundation has dedicated itself in large part to the development and sustenance of teaching and teachers. It established the TIAA/CREF retirement funds.[1] It currently has a massive program to evaluate graduate education nationwide. It sponsored the book [GTM]. One of the contributions of the Carnegie Foundation has been to advocate a rethinking of the way we evaluate professors and reward them. Another is to consider the service roles of college and university faculty. The Carnegie Foundation also offers grants for a variety of teaching activities.

The first and primary piece of advice about this topic is to take your teaching seriously. You may feel in your heart of hearts that the only thing that *really matters* is research; ever since the advent of NSF grants about fifty-six years ago, that has been the commonly held belief in our profession. But teaching is what pays the bills. It is the most visible thing that we do, and it is the thing that we do *that others understand*. Even the Dean has only the vaguest sense of what your research life is about, but he/she certainly knows about your teaching. Our reputation around campus hinges on our ability to teach. Our credibility with the administration hinges on our ability to teach. So you are doing your department and your colleagues a service to do a (more than) creditable job teaching.

Some teachers are jocular and are always clowning around with their classes. Others are quite serious—nearly morose. Still other teachers make the learning process a group activity; they are more "the guide on the side" than the "sage on the stage". All of these are valid and effective didactic

---

[1]TIAA stands for "Teachers Insurance and Annuity Association" and CREF stands for "College Retirement Equities Fund". These two programs were created by Andrew Carnegie to ensure that America's teachers were well cared for in their golden years.

methods *in the hands of the right individual.* I like to tell a joke now and then to my classes; self-deprecating humor seems to be particularly effective. But I never clown around. I am never morose. I am always the sage on the stage, but I go to great lengths to encourage class participation. I give the *impression* that we are all learning the material together. It makes the students happy and keeps them alert and participating. These are my methods; they work for me. I cannot claim that they would work for everyone.

It is certainly worthwhile to have a teaching mentor chosen from among the senior faculty. Any new teacher is bound to have scads of questions about every aspect of managing a class, writing exams, preparing lectures on tricky topics, handling graphics, using technology in the classroom, and any number of other hot issues. It always helps to consult a more experienced practitioner. It is also useful to brainstorm with your peers—other faculty at your level in the profession—about issues that have come up with your classes. Beginners find it comforting to learn that their colleagues at the same level have many of the same issues and problems. And that they can work them out together.

You do *not* want to develop the reputation in your new department of someone who teaches to the exclusion of all else—hangs out with the students night and day, spends untold hours preparing extra lessons and handouts, and so forth. Quite frankly, behavior such as this makes it appear that you have no perspective on the job, that perhaps you are immature, and that you are simply reveling in the puerile pleasures of hanging out with eighteen-year-old kids. Always keep in mind that teaching is a *very important part of what you do,* but it is *not* the only thing. When tenure time comes around, you will be evaluated for a variety of characteristics—*not just teaching.* They need to all be in place.

Colleges and universities these days take teaching seriously. This is in part for sociological reasons, and in part because institutions are in fierce competition for good students. Also good teaching makes trustees happy, makes alumni happy, and makes donors happy. Finally, teaching is the most visible thing that the math department does. Our reputation around campus hinges on our ability to teach, and on our reputations as teachers. The Dean evaluates the math department (and all departments) on a variety of characteristics; but teaching is one of the primary ones, and one that he/she emphasizes in his/her meetings with the Chair. The Dean's view is that teaching is our "main product". So it is good business to attend to it. See [BRU] for a contemporary and timely take on the significance of teaching in our world today.

Although books like [KRA1] contain considerable detail about all aspects
of the teaching process, it is not out of place to outline some of the key
points here. The main message is to take the process seriously. Imagine that
the members of your class include the Pope, your mother, the department
Chair, and your best friend. Of course you will therefore want to prepare
your lectures/lessons carefully. You will want to make them interesting. You
will want to impress people with your erudition and your wit. This does *not*
mean that you want to show off and ham it up. It *does* mean that you want
to make the best and most serious possible effort.

All the additional advice that I shall tender below will follow naturally
from the main points made in the last paragraph. It is all just good, common
sense. But sometimes the obvious must be enunciated:

- Write out a good set of notes for each class/lecture. You may never
  actually refer to these notes during the class period. But the very fact
  that you took the time to prepare them means that you were mentally
  ready for this presentation, you were on top of all the details, and you
  carried the job off with professionalism and dispatch.

- Be prepared to answer student questions. After all, why should students bother to come to class if they cannot ask questions and interact
  with the instructor? Make them feel comfortable in the process. Better
  still, make the students feel as though they are contributing something
  to the learning process. At the end of the class hour, everyone should
  feel good about what took place and what everyone's role was in the
  interchange.

- Give fair and well-thought-out homework assignments. Some textbooks
  make it easy to pick out a good distribution of drill exercises and a nice
  selection of "thought problems". Others make it practically impossible.
  But it is your job to cut through the thicket and cook up a homework
  set that is **(i)** of the right length, **(ii)** of the right level of difficulty, and
  (iii) with the right amount of material. Bear in mind that somebody
  is going to have to grade this exercise set, and somebody is going to
  have to make sense of it after it is graded. This is *not* a trivial task.
  It can really affect the quality of your relationship with your students.
  Take it seriously.

- Meet your office hours and be ready to interact meaningfully with students. Generally speaking, students will not flock to your office hour. But if word gets around that you are there and you are helpful and you treat students with the courtesy and decorum that suit the situation, then word will spread. Students will grow to depend upon you—and even to like you. That is probably a good thing.

- Let students know that you care about the class and you care about them. You do not have to say it; students will be able to tell instinctively. Show up for class bright-eyed and bushy-tailed and ready for a good, instructive time. Act like a human being. Respond to students enthusiastically and with sincerity.

- Give fair and well-thought-out exams. Let me assure you that students take exams *very* seriously. So respect that fact of life and respond in kind. A good exam will send students away thinking that they've had a fine educational experience. They will feel good about the class and good about themselves. A bad exam will make students feel like committing ritual suttee. A good exam has the right number of questions of the right length. Students should feel that they have time to work each problem and to check it. They should not feel that they are rushing at breakneck speed to get to the end. The assignment of points to each problem should be fair and well-reasoned. The grading of the exam should be even-handed and realistic.

  You want to give an exam with the property that the mean is sixty-five or seventy (out of one hundred). That way you can easily curve the scores and assign *A*s, *B*s, *C*s, and so forth in a fair and logical manner. What you *don't* want is to have to stand in front of a room full of hostile students and explain why the fact that the mean on this exam was thirteen is not too important. If you square your score and divide by the ERA of the second pitcher in the last Yankees/Dodgers game, then you will get a more realistic estimation of how you are doing in the class. This last is a surefire recipe for unhappiness and possibly open rebellion. A little care and forethought can stem such troubles.

- It will certainly happen, during the course of the term or the semester, that a student will come to you with a problem that is out of the ordinary flow of events in the course. Perhaps the student bought a plane ticket for the December holidays that conflicts with the final exam. Perhaps the student has some personal problem that has prevented him/her from doing the last two homework assignments. Or it could be something even more sticky. The easy way out for you in such a circumstance is just to take the hard line: Tell the student that it is his/her responsibility to get the work done and to meet the exams; failure to do so will be punished accordingly. The perhaps more humane thing is to try to work with the student and find a mutually agreeable solution. In the case of the interfering plane ticket, schedule the exam for this particular student at another time. In the case of the missed homework assignments, cut the student some slack; surely he/she will learn something by actually completing these assignments. For nasty personal problems, you probably should not get involved. You can, if appropriate, refer the student to an appropriate counselor or Dean.

- If you can arrange it, and if it doesn't interfere with the next user of your classroom, then arrange to hang out in the front of the room for a few or several minutes after each lecture. Students feel very comfortable coming to you with queries or quips or observations as they shuffle out the door (by contrast, students are often rather timid about coming to your office hour). You can get a true sense of how the class is going, how satisfied the students are, how the learning flow is working, by interacting with students in this semi-informal setting. A friend of mine once joked that the secret of success in undergraduate teaching is to never let a student get between you and the door. That's a suitable *bon mot* over coffee with your friends; but it is not the advice that you should follow in practice.

- You want students to think of you as their ally and facilitator in conquering a body of new material. Students should think of you in a friendly and constructive manner. This does not mean that you should *literally be their friends*: You should not be going to bars with them, or having parties with them, or playing ultimate Frisbee with them. But your relationship with your students should be very positive. They should *want* to come to class, and *want* to see you in your office hour. Conversely, you should *want* to meet your class three days a week; it should be something you look forward to as much as your tennis game.

2.2 How Not to Teach

There is a certain cachet to looking down on all teaching activities. If you are in a large math department, you will have no trouble locating a subset of its denizens who sit around bad-mouthing the students, bad-mouthing the teaching assignments, bad-mouthing the university administration, and generally painting a negative and miserable picture of the entire teaching enterprise. This group always welcomes new members, and you will have no trouble making a new group of friends rather quickly.

Well, it's fine to be friends with these people. As a general rule, you should be friendly with everyone. But I would caution you against becoming an active member of this group. First of all, their mission is not constructive. Secondly, you want to make it your goal to be a good—if not an outstanding—teacher. Third, you don't want the powers-that-be in the department to associate you with such negativity.

Like it or not, the welfare of the math department on campus, the way it is perceived and the way it is funded, depends decisively on how math department *teaching* is perceived. If people think that the mathematicians are a bunch of out-to-lunch eggheads who can't teach ice to Eskimos (and often this is precisely how we are viewed, although usually unjustifiably so), then the consequences will be grim.

You may be a crusader, and want to convince all of your colleagues to be teaching heroes. This is probably not realistic. You cannot—especially if you are a new, young faculty member—turn the entire math department around and transform everyone into a model teacher and citizen. What you *can* do is to carry your own weight, be a role model for students and for other faculty, and show people that you are a team player who does his/her job to the best of his/her ability.

I have written in detail about the nuts and bolts of good (and bad) teaching in other places (see particularly [KRA1]). I shall not repeat all those details here. Suffice it to say for now that you should handle each of your teaching assignments with suitable dedication and care. Write a good syllabus. Prepare each lecture or classroom lesson carefully and in detail. Practice your delivery and make it shine. Get a colleague or friend to watch you teach and make constructive remarks. It is easy to arrange, through the Teaching Center, to have yourself videotaped. You will find this to be a revealing and (on occasion) terrifying experience. But it will tell you a lot about how you come across.

Write neat, clear, coherent assignments and exams. Make yourself available to students. Be fair, reliable, and punctual. Students need to know that they can depend on their instructor, and that the instructor will not surprise them in nasty ways or pull the rug out from under them when least expected.[2] In this manner, if you are like a good parent to your students, then they will respect you and forgive some of your other shortcomings.

It is quite easy to talk yourself into a downer over your teaching. While it should be an uplifting and inspiring part of your life, it *could* turn into a distinctly negative experience. Especially if it eats up too much of your time, if the students turn out to be more of a pain than a pleasure, if your classes offer you more aggravations than rewards, or if you feel that you are not getting through to the students and instead are just wasting your time.

Looking back on the hundreds of classes that I've taught, it is easy to see how any of them could have gone south. I could have had an extremely unruly and uncooperative class. (The fact is that it only takes a few troublemakers to turn a whole class bad.) I could have been consistently unprepared, made too many mistakes, given bad exams, and generally created a bad morale situation. I could have been insensitive to the students' level, the students' needs, the students' backgrounds and goals.

If you look over the preceding paragraph, you will see a few items that are (in principle) the students' fault and several items that are (indeed) the instructor's fault. This is no accident. When you teach a class, *you* are in charge. It is your one-person show. You set the tone, you shape the class and turn it (we hope) into an effective working unit, and you see it through to the end. There are a good many mistakes that you could make to prevent this from happening effectively, or at all.

The principal guiding rule when you teach a class is to treat people the way that you want to be treated. This precept entails preparing your lessons well, going to class with a good attitude, treating everyone with dignity and respect, going out of your way to be helpful, and exerting every effort to be fair and evenhanded. None of these features should require a special effort on your part; if you are a committed teacher, then they should all come rather naturally. If instead the attitude that you bring to your teaching is that of a gang member on a street corner, then you will reap the predictable

---

[2]This is a lot like being a good parent. Children need to know that they live in a dependable and stable household, and that their parents will always be there for them. Students are much the same.

(negative) reward.

I have learned a lot from watching Jay Leno and David Letterman on television. Of course both those guys are very funny—which I am not. But what is special about them from my point of view is that they are *really good* at rolling with the punches. No matter what anybody says or does on their shows, these hosts know just what to do to turn the situation to their advantage. Everyone comes away feeling good, and as though he/she has contributed something positive to the occasion. And that is just the atmosphere that you should foster in your class. If a student asks a stupid question, or accidentally dumps all his/her books on the floor to create a disruption, or has a pizza delivered right in the middle of your brilliant lecture on the mean value theorem, or if someone's cell phone goes off during your recitation on contour integration, turn the situation into a plus. Crack a joke, or make a witty observation, or have a piece of the pizza yourself. *Do not* endeavor to turn the transaction into some kind of morality play in which you are God meting out eternal punishment. That is a sure way to alienate everyone in the room and turn the class against you—not just for that day but for the rest of the term.

It is really a privilege to stand up in front of a group of young adults for the duration of a semester or term and dole out your aggregate wisdom. You can make the experience a real pleasure if you think about it in human terms. You certainly have the math under control—that is what your Ph.D. attests to. Now you must get the human side of the picture into focus.

Don't be afraid to ask a senior mentor in your department for tips about teaching. Every school has its own special needs, special features of the students, particular quirks of the curriculum. A seasoned veteran can help to acquaint you with some of the local wrinkles, and to come to grips with the teaching life in your new department. This senior mentor will know how teaching is judged *in that particular department*, and will be able to guide you in developing the right skills and the right values.[3]

Don't hesitate to consult the Chair, or the Vice-Chair for Undergraduate Affairs, or the Coordinator of Lower Division Teaching, for direction and

---

[3]When I visited a particular math department in Australia, the Chair told me that the freshmen at that school were all hooligans. They ran up and down the aisles, shouted epithets in class, threw paper airplanes, and generally treated the classroom experience like a brawl. He attributed this idiotic behavior to the fact that most of them still lived at home, so they were highly immature. But the bottom line was that his new faculty had to receive special training, and extra counseling, in how to handle these classes.

advice in dealing with particular teaching questions or broader philosophical issues. After all, this is your life that you are shaping. Your research is directed to a very limited and recondite audience. But your teaching reputation will be known to everyone.

Scholars from other countries sometimes have trouble adjusting to our teaching system (see Section 5.6). The nature of student preparation, and the way that the curriculum is structured, is quite different in Italy from that in the United States. Often a well-meaning instructor from Europe will stand in front of a calculus class pitching the material as though he/she were teaching real analysis to juniors. Obviously this will not work. If you are a new instructor here (in the U.S.), you will have to take special pains to acquaint yourself with the curriculum and the students at your new school. Choose a respected teacher/colleague—preferably an American—and sit in several of his/her classes. Best if this is a calculus or other elementary class, so that you can acquaint yourself with the ebb and flow of instruction in such a setting.

I can tell you that the department and the university administrations will be paying particular attention to *your* teaching as they assess your advancement in the system. You will have to work especially hard to make your way through this unfamiliar territory.

2.3   How Teaching Is Evaluated

Saunders Mac Lane used to say frequently—both in print and in public utterances—that we all know quite well how to recognize and evaluate and reward good research, but we have no idea how to recognize good teaching. Teaching is more subjective. It should be left to the individual.

Whatever the merits of Mac Lane's arguments, they ignore the realities of the world we live in. Universities today are in intense competition for the best students. Quality teaching speaks to the students and, perhaps more importantly, speaks to their parents. Deans and Provosts and Chancellors and, yes, even Boards of Trustees care deeply about teaching. If you want to get tenure, if you want to get promoted, if you want to get good raises, then you had better learn how to teach well. And I don't just mean well enough so that you can go home each evening with a clear conscience. While most of the time you can get by with a clean teaching record that is better than average, it is becoming more and more the case that the administration wants to be able to say that you are really a special teacher. I mean good enough so that the students sing your praises, so that it is *demonstrable* that you are an outstanding teacher. At my own university, this is a necessary condition for promotion to full Professor. The university's view is that, by this point, you've had plenty of years to practice the craft. Now you should have it right. Other universities may not be quite so passionate, and more inclined to pay lip service to the whole business. But there is no question that you are better off being a very good teacher than not.

Surely the most common method for evaluating teaching is with student teaching evaluation questionnaires. The traditional method for handling this device was for the instructor to hand out hard copy teaching evaluation forms at the end of a class period towards the last days of the term. The students were to sit right there and fill them out. And a student volunteer was to collect them and deliver them to a departmental secretary (so that the instructor could not see them right away and be unduly influenced in his/her subsequent grading of the course). Even this simple device has undergone some development. At my own university right now teaching evaluation is done OnLine. This has the advantage that there is no longer any question of fairness, or of the instructor seeing the evaluations before he/she should.[4] And the results are compiled automatically. One disadvantage is that the instructor is no longer distributing the forms *in class* and asking students to complete them *on the spot*. So the response rate is now much lower. Again, at my own university, the entire student evaluation system for teaching used to be handled by the student body organization. There was nothing wrong with this, and the system seemed relatively fair and evenhanded. But it made some faculty nervous to think that a system on which so much of our welfare (promotions, raises, etc.) depended was not objectively institutionalized. Now, with the OnLine system, the teaching evaluation mechanism is about as objective as you could want.

There *are* other methods for evaluating teaching. When a candidate is up for tenure or promotion, it is quite common for the Chair to send a couple of faculty to observe the person's teaching firsthand. The evaluator writes a few paragraphs describing what he/she saw. This technique can be quite useful, as faculty evaluators aren't worried about a grade, or about their ability to

---

[4]When I taught at UCLA an Associate Professor of Social Studies got in a *heap* of trouble for *forging* (i.e., manufacturing from scratch) his teaching evaluations. His teaching assistant outed him. Seems that members of the Social Studies faculty had been under considerable pressure to beef up their teaching reports. This was his solution.

understand the material. So presumably they can be more objective. And they speak, of course, from greater experience.

I have always thought that an ideal way to handle teaching evaluation would be to have a professional psychologist sit down and talk to each student privately. Such a person would know how to draw the student out, how to ask the right questions, and how to elicit the information that was truly desired. Unfortunately, such a system would be quite expensive and time-consuming, so it is rarely used.

Many schools now have midterm teaching evaluations, and these can be quite useful. These are just between the students and the instructor (usually), and no records are kept. But they are a way for the instructor to find out—in an objective sense—how the course is going, how students are responding to the material, and what can be improved. Just so, many schools now have midterm grades for students. You would think that a student could tell that he/she was flunking a course. But often he/she cannot, or simply cannot take off the blinders to see. It is just the same with instructors. It is far too easy for us to convince ourselves that we are doing a creditable job when in fact there is plenty of room for improvement. *Objective evaluation* is the way to get your hands on the necessary information.

One thing that I should stress—and this is an empirically verifiable fact—is that *self-evaluation* is the least reliable form of assessment. *You* simply cannot tell whether you are a good and effective teacher. For one thing, you are not objective.[5] For another thing, you are already a master of the material. And, lastly, you have no way of telling how well your presentations and lessons are being received. The long and the short of it is that, if you really want to develop and improve your teaching, then you must get student and third-party input.

2.4   How to Establish a Teaching Reputation

If you are really an outstanding teacher—the sort of person whom students talk about and recommend to their friends—then the word will spread. If you are not yet tenured, then the Chair and the Executive Committee will be

---

[5]Almost everyone I know thinks that he/she is a great teacher. These people can stare at a handful of *really* negative teaching evaluations and say, "The students don't really know what they are talking about. They wouldn't know a good teacher if they saw one. I am *really* a talented pedant. Just look at all the great stuff I show them."

monitoring your teaching dossier. They will figure out that you are someone special. It would not be at all surprising if the Chair got some unsolicited letters from students praising your abilities as a teacher and mentor. This will be a great plus for your dossier.

Of course if you are an active participant in a seminar, then your colleagues will have firsthand experience of your lecturing and expository skills. And that will lead them to conclude that you know how to teach.

If you really want to establish your reputation, then there are more creative things that you can do. You could start a teaching seminar. Most likely there are a number of people in your department who care about teaching issues—calculus reform, group learning, self-discovery, educational labs, and the like—and want to discuss and develop them. Your contribution here could be a very positive part of departmental life. Just be careful to run the seminar as an intelligent exchange of information and ideas. You do nobody any good if you try to come across as the know-it-all on teaching.

You could volunteer to be on the Undergraduate Committee, and get involved in ongoing teaching and curricular projects in the department. Once you are more senior, you could volunteer to serve a stint as Vice-Chair for Undergraduate Studies. Or perhaps as Coordinator of the Lower Division Curriculum (although this could be a staff position, and not for you).

You should be aware as you do these things that you will garner the reputation that you deserve. If you spend all your time on teaching activities—which is certainly a valid and worthwhile thing to do—then your colleagues will peg you as a "teaching type". This is something you can be proud of, and you need not shirk it. But one corollary could be that, when there are discussions of research or hiring issues, then your opinion will not carry as much weight as it once did.

I certainly know perfectly fine mathematicians—quite a number of them, in fact—who at a certain point in their careers said to themselves, "Things have changed. I no longer have an NSF research grant. I don't receive the speaking invitations that I once enjoyed. The invitations to conferences are fewer and further between. And I now have a more mature perspective. I feel that I've made my research contribution. I have fifty papers on MathSciNet, and that is an adequate output for a lifetime. I have now developed an interest in teaching and administrative issues, and that is how I am going to spend my time." This is great, and is certainly a thoughtful and (potentially) productive position to take. It beats the heck out of spending your time staring at the wall wailing "Woe is me". There is plenty of grant money available for teaching and administrative activities (probably more than for research in pure mathematics). And your institution will appreciate your efforts.

For the fact is that many, if not most, university administrations take an attitude quite similar to the one just enunciated for an individual mathematician. When you are working your way up through the ranks, you are supposed to be a fire-in-the-guts researcher. You should be obsessed by the research life. But, as you grow older, your focus will change. The institution will ask more service from you. Your interests will broaden. You will realize that teaching and research coexist peacefully, and support each other. You will be able to put an emphasis on *both* aspects of your efforts. As a result, you will want to make different types of contributions. Everyone will respect that decision, and many will appreciate it.

The teaching life is a good life, and you should give it its due. You are much better off having a positive teaching reputation than the opposite. Teaching and research are *not* disjoint activities, and you should not perceive them as such.

2.5   Teaching in Large Lectures

Teaching calculus and other meat-and-potatoes courses in the large-lecture format is just a fact of life these days. Given how many faculty there are in a typical math department, given how many students take calculus and precalculus (and the myriad other elementary courses that we offer), and given the teaching load that we all expect and deserve, there isn't any way to cover the teaching obligation without having lone faculty standing off against an auditorium filled with three hundred less-than-enthusiastic students.

It is not the best possible circumstance in which to teach. In fact I wouldn't even rank it in the top ten. It is a trial for all concerned. Students have a hard time getting anything out of the class. The instructor has a tough time communicating—even maintaining order. Likely as not the instructor will not be able to collect and grade homework, thus increasing the already built-in alienation.

Teaching heroes like Ole Hald of Berkeley (see Section 7.7) have ways to bring life into this otherwise dreary scenario. There is a real craft to getting along with a big class, and making the experience at least tolerable for all concerned. Some instructors will appoint ombudsmen to open up the lines of communication between the professor and the students. Other faculty will have extra office hours. Still others will give extra review sessions.

Of course the large lecture is augmented by a collection of recitation sections taught (typically) by graduate student TAs. The typical morale in these recitation sections is not terrific either. Students feel (usually unfairly) that they are being taught by second-rate instructors at best, attendance is poor, and discipline is a problem.

The book [KRA1] gives considerable detail about how to handle this type of less-than-ideal teaching situation. When you were young and dreaming of being a tenured faculty member at a good college or university, you surely did not fantasize about teaching an unruly class of three hundred. But that is the reality of life. You need to get used to it. More than that, you need to deal with it effectively—even get good at it.

2.6 Choosing a Textbook

Selecting a textbook for your course can be a pleasure or a pain. If you are teaching calculus, then there are a good many choices. And there are interesting differences among the different calculus books. Some are written in the teaching reform style and some are in the traditional style. Some emphasize computing and some emphasize applications. There is even a calculus book that is in the style of a comic book (see [SWJ]). If you are teaching a course on Arakelov theory, then there are only one or two books (see, for instance, [LAN]), and they are quite recondite. Few people can read them. For a typical undergraduate course there are myriad choices, and you will be doing yourself a favor to choose carefully.

Many courses fall in between. There will only be a handful of texts, and you may not like any of them. Of course that would be a good excuse to write your own text (see Section 5.1). But that road is not for everyone.[6] You could also cobble something together from several textbooks, but be careful.

---

[6]When Walter Rudin was a Moore Instructor in the late 1940s at M.I.T., he was assigned to teach undergraduate real analysis. He quickly realized that *there was no text* (at least not in English, readily available and accessible to American students). Of course this was in the postwar period—a different age. People had been in the habit of learning real analysis—and many other mathematical subjects—from sets of notes that circulated privately. So he decided to write a text. And Walter's effort turned into one of the great classics of twentieth-century mathematical writing. Rudin's *Principles of Mathematical Analysis* later won a Steele Prize.

It is technically illegal, because of copyright law, to make photocopies of parts of several books and put them together as a text for your students—it is especially illegal if you *sell* it to your students.

These days most major publishers have sales representatives, and these charming young people will use all their wiles to convince you to adopt their book. Particularly if you are in charge of the calculus course at a big state university, your choice of text could be a 1,500-unit adoption. That would be a big sale for any calculus rep, and they could offer all sorts of inducements to get you to swing their way.

The book that you choose for a class can really affect the way that the course will go. If the text is full of errors, or if the exercise sets are poorly designed, or if the explanations are confusing and misleading, then you will find yourself spending a disproportionate amount of time compensating for the shortcomings of your textbook. This is counterproductive, and can create a negative morale in your class. I have actually known instructors who have said to their class, "This text is so bad that I am going to recommend that we abandon it and adopt another." This in mid-semester! Of course that creates problems of its own, as many students cannot afford to buy another textbook. Students also find such a change demoralizing and confusing.

So you must put in some legwork to select a good text. Certainly asking experienced colleagues for their advice is a good first step. On the one hand, you cannot actually *read* each textbook that you are considering. You cannot work every exercise in every book. But you can skim significant parts of each book, read certain key sections carefully, sample the exercises for accuracy and consistency, check the examples for clarity and incisiveness.

The lesson here is that you want to be careful and meticulous in adopting a book. You don't really know a text until you've used it for a semester or term, and lived with the examples and the homework problems and the applications. But you can ask a more experienced colleague to recommend a text, and you can consult Internet chat rooms (or even `Amazon`) to read reviews. The *American Mathematical Monthly* carries reviews of many new texts, and `http://www.maa.org` carries many reviews OnLine. See also the *Mathematical Gazette*. The more information you have, the better off you will be.

2.7 Teaching Cooperatively

At a large state university, the following scenario for calculus is quite common: In the fall semester, there are 1,500 students. These are divided into five large lectures of three hundred, each taught by a Professor. Then there are about fifty problems sections, each taught by a graduate teaching assistant (or TA). You can see that this is a situation that requires considerable management and oversight. It is common to put a Professor—usually one of the five who is giving the lectures—in charge of the whole course.

Of course the five Professors will meet regularly—perhaps once per week—to discuss issues connected with the course. These could include

- management of the TAs

- construction and scheduling of the exams

- proctoring of the exams

- grading issues (for both homework and exams)

- pacing of the course, order of topics, reaching of milestones on key dates

- overall grading policies

- coordination of office hours

- coordination with the math lab (if there is one)

This is another opportunity for you to interact with your colleagues in a constructive meeting of minds to tackle a common task. It is really not all that pleasant, but you can make it congenial and productive. After all, you all have a common job and a common goal. It is in your best interest to work together to make this come out as fruitfully as possible.

It is really best if the five lecturers stick to a rigid syllabus and lesson plan, so that on any given day they are all teaching the same thing. That way, if a student misses a class, then he/she can go to a different lecture and not fall behind. Also the lecturers will all get to the same milestones at the same time, so that they are all ready to give just the same midterm. This all requires a bit of discipline, but it is not hard and the payoff is considerable.

There are benefits that you can offer the students in this otherwise trying learning situation. You can tell students that they are welcome at *any* professor's office hour. This will be well-received, and not increase your business appreciably. Students can have a choice of several review sessions to attend when exam time comes around. They can attend several different recitation sessions if they wish to do so.

Typically the five professors in a situation like this will have a common syllabus, give common exams, and coordinate their grading. If this is done right it can save labor for everyone. In addition, you can learn from each other in the process. You should, of course, expend every effort to get along with your colleagues here, to see their point of view, and to work together to make everyone's life easier. Such cooperation is obviously in everyone's best interest.

2.8   Teaching Outside Your Specialty

It is a fact of life that, in many academic disciplines, people stick to their specialty in everything they do. A Professor of English who is an expert on John Milton will only teach courses on Milton. A Professor of Biology who is a geneticist will only teach courses on genetics.

Mathematics is not like that. For one thing, we tend to be more broadly trained in our bailiwick than other academics. For another thing, our curriculum is much more catholic. We only have so many faculty, and the courses need to be taught. So, while it is a special pleasure to teach courses in your research area—and you will likely do so every few years—you will find that you are asked to teach a great variety of courses.

I am a harmonic analyst by training. But I have taught courses in formal logic, linear programming, special functions, abstract algebra, differential geometry, and a variety of other subjects. I have found this to be a good deal of work, but it was enriching and exciting. I was happy for the experience. You, too, should find that this is a broadening aspect of your life, and one that you can look forward to.

2.9   Media

Today it sometimes seems as though media control our lives—and I am not talking here about the CBS Nightly News, but rather about computer and other hardware media. Especially because communication (in both oral and written form) is so important to our profession, we must deal with media.

In the old days, we would present our lectures with chalk at a board and do our writing with a pen on paper. If the writing was to be ultimately published, then *someone else* would type it up, typeset it, copyedit it, and get it into print. Today many aspects of communication have changed dramatically. Many lectures are given with `PowerPoint`® or some variant thereof. Today many of the publishing functions are performed by the author using electronic methodology. It is a visceral change that has impacted all our lives.

In the old days, if you gave a lecture (with chalk on a board) and somebody missed it, then your response was likely to be, "Sorry you missed my lecture. Maybe you can catch it another time. I could give you a copy of my notes if you wish." Of course, likely as not, the notes only gave a sketchy preview (in semilegible form) of what the lecture really was about. And the notes would not include any graphics. So the end of the story is that the person who missed the lecture indeed missed out and won't be able to make it up (although perhaps he/she could borrow another student's notes).

The next step in the evolution of presentation technology was to use either an opaque projector (or epidioscope) or an overhead projector. Here an opaque projector projects (onto a screen) from a printed page or book. An overhead projector projects from transparent slides. Advantages of these media are several:

- Your materials are displayed on a large screen, hence visible to a large audience.

- Complicated graphics and tedious tables of data can be prepared in advance on slides.

- Everything looks (in principle) quite polished.

- You will have your presentation in your archive to use on another occasion.

- You can share your lecture materials with people who missed the formal presentation.

There are also several disadvantages:

- It requires considerable extra time to prepare slides in advance.

- The use of a machine like this puts a psychological barrier between yourself and your audience. Some of the intimacy of a chalkboard presentation is lost.

- There is a temptation to put too much material on each slide. You must learn to design slides effectively.

- You have to retrain yourself to pace a slide-driven lecture properly. Since everything is written out in advance, there is a great temptation to go too quickly.

- The use of slides tends to lock you into a linear order of presentation. With chalk at a board, you have more control. You can jump around, restructure the talk in response to a question from the audience, or pursue a digression. With slides this is more difficult. As a result, a slide presentation *could* be less lively.

The truly modern way to prepare a presentation is with `PowerPoint`. This Microsoft product creates a computer file with individual frames to project onto a screen. You cycle through the frames by hitting the right-arrow key on the computer keyboard, or by using a handheld remote. `PowerPoint` is quite a flexible and powerful tool. It allows you to include color, videos, animated graphics, sound, and many other media devices. Certainly for business presentations it is often the way to go. When I was Chair of my department, I always felt that a presentation to the Dean should be done in `PowerPoint`.

Some people are critical of `PowerPoint` (see, for instance, [TUF1], [TUF2]). It can be too slick, it can distance you from your audience, it can be superficial. But, like any tool, it can be effective when used well and in the proper context.

It is tricky to incorporate sophisticated mathematics into `PowerPoint`. Some mathematicians prefer to use the freeware product `Beamer`.® The German creation `Beamer` is a LaTeX package that produces a `*.pdf` file that has individual frames, just like `PowerPoint`. And `Beamer` has much of the functionality of `PowerPoint`—animated graphics, sound, color, and so forth—but it also allows you to include any mathematics that can be rendered in TeX.

Of course one of the great advantages to preparing your lectures electronically is that they are then quite portable. You can put them on the Web, you can send them to friends, you can even publish them. You still have to worry—quite a lot, actually—about the design of each frame, and about your pacing. But this is part of today's world, and if you are interested, then you will do it.

Of course publishing these days is a whole new ball game. When I was an Assistant Professor at UCLA, the Department had *three* full-time manuscript typists. I would write my latest research paper by hand in ink on paper and humbly submit it to the crew. One of them would labor away—for up to a week!—to type up the manuscript. This was done on an IBM Selectric® typewriter—using special type balls (or *elements*) for all the special symbols. And then the project would be given to me for proofing. An indication of how primitive things were then (around 1975), and how sophisticated we are now, is the *form* in which the proofs were given to me.

You would think that the typist would have given me a photocopy of my paper to proofread, and that he/she would retain the original typescript for safekeeping. Thus I would make my edits and corrections in red ink on the photocopy. But *no.* In point of fact the typist would give me *the original typescript* with a translucent overlay clipped to each sheet. Then I was provided with a special red wax pen to mark the overlays with my corrections.[7] You see, in those days photocopying was still considered to be something of a luxury. So the overlay system was UCLA's attempt at economy.

Of course now the world has changed dramatically. Most departments no longer have manuscript typists. The majority of mathematicians prepare their own manuscripts—*even book manuscripts*—in TeX. Advantages of this new system are

- The author now has complete control over layout, content, and accuracy.

- The author can, if he/she wishes, provide graphics using `Adobe Illustrator`® or `Corel DRAW`®or `xfig`.

- The author can use a mathematical utility such as `Mathematica`® or `Maple`® or `MatLab`® to create and render and format graphics.

---

[7] You can imagine that some faculty, not being gifted at following instructions, would lift the overlays and mark with the wax pen on the original typed pages. This did *not* please the typists.

- The author retains the electronic `*.tex` and `*.dvi` files and can post them (or a derivative file such as a `*.pdf` file) on the Web.

- The author can submit his/her paper directly to a journal or a book publisher by either sending the `*.pdf` file as an *e*-mail attachment or posting it on a Website. Many times people will first submit their work to the preprint server `arXiv`. In that case submission is performed in TEX. With many journals you can submit a paper just by providing a pointer to the `arXiv` posting.

There are also disadvantages to the new system. Some of these are:

- The mathematician is now required—for manuscript preparation—to learn skills (TEX, `Adobe Illustrator`, various operating system maneuvers) that used to be relegated to staff.

- The mathematician must learn various operating system commands and techniques in order to be able to manipulate TEX source, graphics, and other files to get the desired (unified) result.

- The mathematician must spend a *lot* of time typing.

- The mathematician must maintain an electronic profile OnLine.

- The mathematician may have to learn `HTML` and how to maintain a Web page.[8]

This last point is perhaps worth some discussion. These days most of us have a personal Web page. That page commonly includes:

- a version of the CV

- courses taught

- professional affiliations

- national committee service

- conferences being organized

---

[8]One can easily imagine that André Weil or Carl Ludwig Siegel would have objected strenuously to learning any of these things.

- links to newsgroups and chat rooms

- a list of publications and preprints

- links to those publications and preprints

This new facet of life has many pluses:[9] It is now much easier to keep up with what other mathematicians are doing. If you want to get a new preprint, it is quite natural to go to the author's Web page and grab it. But a negative is that the progenitor must spend time maintaining his/her Web page. Ideally one should update the Web page once per week. One should update the OnLine CV once every few months. And one should keep the posted preprints and papers current. It's a lot of extra work, and time-consuming as well.

2.10 Research

The role of research in your life is discussed in detail in Chapter 4. But let me say point blank that, if you are at a research university—what we in the trade call a Group I school[10]—then research is supposed to be a big part of your life. It would not be at all unreasonable if half of your professional time were spent on research—either reading, or engaging in professional correspondence, or thinking, or calculating, or going to conferences or seminars, or applying for grants, or writing. At other schools, where there is a greater emphasis

---

[9]Many mathematicians, indeed many people, have an extensive component of their Web page that is about their personal lives. There can be photos of the family, photos of hiking trips, material about the offspring's virtuosity on the violin, pages about the family pets, extensive philosophical musings or blogs, and so forth. My personal view is that your math department Web page is a business document, and should be treated and developed as such. Others may differ.

[10]Pure math departments are classified as Group I, Group II, Group III; these are departments in the United States having doctoral programs in mathematics. Group IV contains U.S. departments or programs of statistics, biostatistics, and biometrics reporting doctoral programs. Group V contains U.S. departments in applied mathematics, applied science, operations research, or management science which have doctoral programs. Group M lists U.S. departments which grant Master's degrees as the highest degree. Group B lists U.S. departments which grant the baccalaureate degree as the highest degree. The publication [GMF] gives a numerical ranking to each mathematics department in this country. The highest score is 5.00. Those departments with a score in the 3.00–5.00 range are classified as Group I. Those with a score in the 2.00–2.99 range are classified as Group II. All other pure math departments are classified as Group III. There are presently forty-eight Group I departments and fifty-six Group II departments.

on teaching and service, the amount of time spent on research will be less. Certainly to get tenure you will have to have a track record in scholarly publishing. As you rise through the ranks you will find that more and more of your time is taken by departmental work and teaching duties, and your scholarly life will take a back seat to those activities.

If you work at an industrial job, or at a government research facility, then the only thing that you will do is "research". And some administration (which will grow as you become more senior). Generally speaking, it will not be research of the academic sort. Such operations often work on government contracts, and there are very specific problems to solve or technologies to develop.[11] You will still be developing new ideas, and charting new paths. But it will often be security work so that you cannot share it or publish it or get credit for it. And often you will be applying *known* mathematical ideas to technological questions in a new way. You will *not* necessarily be developing new mathematics as such.

2.11   Committee Service

Many departments will reduce the service duties of a young (nontenured) faculty member. The view is that an Assistant Professor should be devoting his/her nonteaching energies to developing a research program and getting grants. Committee duties are of less importance, and can be carried by the senior faculty. But some math departments pay more attention to infrastructure, and will want to get you involved in departmental life right away. And you will then have a fair portion of committee work.

Take this work seriously. Show up punctually for meetings and be prepared. If there is reading material for a meeting, be sure to read it and understand it. Ask questions of someone if you don't. It is entirely possible that a handout distributed for a meeting presupposes some knowledge of departmental history or of past policies. Of course you have no way of knowing these things. You will have to ask someone if you are to participate

---

[11]A notable exception to this paradigm is the Microsoft Mathematics Theory Group. This remarkable organization in Redmond, Washington and Mountain View, California employs quite a number of distinguished mathematicians, and they are given free rein to study whatever they wish (provided it conforms to certain broad guidelines of areas that are of interest to Microsoft). And they can publish anything they like. This is Bill Gates's gift to the mathematics world.

meaningfully in the discussions.

Often committee work will involve homework. You may be asked to re-search a situation—visit with members of the Physics Department and find out about certain calculus needs, for example—and then write a brief re-port. Take these duties—even though they may seem tedious and tiresome—seriously. It is actually fun to become acquainted with members of other departments. And you could learn something useful in the process.

Try to stay awake during committee meetings. Try to participate mean-ingfully. Say things that are intelligent and useful. If you have nothing useful to say, then have the good sense to shut up.

You will probably not, as an Assistant Professor, be asked to chair a committee. Associate Professors are often asked to chair committees. Full Professors carry much of the burden here. If you are asked, then you should agree to serve and you should apply yourself to the task with aplomb and determination. Do the necessary background reading. Talk to people and find out the history that has led to the formation of this committee. Get a clear picture of what the goals of the committee are and how those goals might be met. Formulate a plan of action. Talk to the Chair and find out what he/she wants this committee to accomplish. Assemble an agenda for this committee. Meet individually with the members of the committee and generate a working esprit de corps. Be prepared to guide this committee to a productive and useful culmination of its work. Write a good report informing the Chair and the department of what was accomplished.

Your service on committees will be remembered, and will give your col-leagues a sense of who you are and what you contribute to the general welfare. Your service on the Textbook Committee, the Tenure and Promotion Com-mittee, or the Colloquium Committee could have a lasting and salubrious effect, and a lingering impact. It will certainly leave an impression. You want that impression to be positive and lasting. It is not that you want to be repeatedly saddled with tiresome committee duties. Rather, you want to be respected and admired and valued as a colleague.

Some committee service can have a lasting and positive impact on your life. For example, serving on the Colloquium Committee will bring you into contact with a number of prominent mathematicians, and will help you to network in the profession. You will often end up taking these folks out to lunch or dinner, and you will get to know some of them fairly well. This could lead to some new invitations—or even new collaborations—*for you.*

It might be mentioned that "service" is a broad concept. As you become more senior in the profession, you may want to consider service to the nonacademic community. Often the local schools need help with their curriculum, or in choosing a textbook. Many times a government agency wants some help with a report, or with some analysis. I think it is important for mathematicians to play a positive role in society, and I hope that you agree.

Service to the profession is also important. The American Mathematical Society, the Mathematical Association of America, and many other professional organizations have significant committees that play a key role in our lives. You may want to be a part of this process. The matter is discussed in more detail below.

2.12   A Panorama of Committees

A math department typically will have quite a selection of committees. Most of these serve a useful purpose, but many of these also exist in part in an effort to create the effect of inclusiveness. Math department members want to feel that they are taking part in the operation. This doesn't necessarily mean that they want to put in the time; but they want to have a say.[12]

Anyway, a typical panoply of departmental committees looks something like this:

- the Undergraduate Committee

- the Graduate Committee

- the Hiring Committee (there could be several of these)

- the Postdoc Committee

- the Endowed Lecture Series Committee (there will be several of these, one for each series)

---

[12]There is one major university in Texas where the Chair twenty-five years ago did everything himself. He had no committees at all. To give him credit, he became Chair at a time when the department was still developing. It hadn't really hit the top tier yet. And he did a lot to bring it forth; there was never such a period of creative and hard-hitting hiring as under this man's stewardship. But I was visiting that department not long after he stepped down. I was astonished to hear people gleefully say, "Gee, I'd like to go for a beer with you, but I've got to go serve on the chalk committee." People had really missed their committee service, and now they were reveling in it.

- the Library Committee

- the Executive Committee

- the Calculus Committee

- the Computer Committee

- the Committee for the Major

- the Colloquium Committee

- the Committee for Faculty Raises

- the Tenure and Promotion Committee

- the Teaching Committee

- the Statistics Committee (assuming that there is a statistics group in the math department)

Of course the Chair will form additional committees and task forces as the need arises. It is important not to have too many committees, as who will serve on them? Nobody wants to be part of more than two or at most three committees. The Chair will of course be an ex officio member of all committees. He/she need not attend all the meetings, but he/she has the right to do so if he/she wishes.

Of course all the committee chairpersons need to report to the department Chair regularly. This is how the Chair delegates duties and gets things done. Sometimes a committee or task force will be assigned a specific task and really not accomplish it. The Chair can then check that off in his/her book as "attempted". Surely some things will have been learned from the exercise, and perhaps more can be accomplished the next time around.

2.13 University-Wide Committees

Part of your job as a university professor is to serve on committees that are *not* subsets of the math department. These are sometimes called university-wide committees. Examples of such committees are

- the Library Advisory Committee

- the Parking Committee

- the Science Curriculum Committee

- the Bookstore Advisory Committee

- the Teaching Awards Committee

- the Tenure and Promotion Committee

Actually, the number of committees at a modern college or university can be quite mind-boggling. You can hardly be aware of all of them. But you are expected to do your part. Especially for promotion to full Professor, you will be expected to have carried your weight in university service. This type of committee work is a standard way to fulfill that role.

If you are really well-known around campus, then you could be asked straight out (by someone in the administration) to serve on certain committees. For most of us, university-wide committee service is arranged through the department Chair, or perhaps through the Dean. My advice would be to serve on at most one or two such committees in any given year. If you are asked to serve on more, the shrewd thing to do is to simply point out that your plate is already full and they should ask someone else.

Of course a committee is a committee is a committee, and you will see that university-wide committees have many of the same characteristics as departmental committees. But there will be differences as well. Different departments have different agendas, and different academic modus operandi, and different value systems. I once served on the library advisory committee and the question arose of whether new journals should circulate. A quite distinguished Professor of German Literature on this committee said, "If a new article in my subject area comes out, then I can just stroll to the library and scan the article in twenty minutes. I don't need to check it out." I was flabbergasted. Imagine an important new article on your stuff appearing in *Acta Mathematica.* Could you stroll to the library and scan it in twenty minutes? You probably could not even figure out what it was *about* in twenty minutes. I think that Herr Doktor Professor could see the puzzlement on my face, so he said, "How is it with you, Steve?" My response was, "If a really good new article in my field comes out, then I might want to spend a year with it. So it is useful for me to be able to check out new journals." I can tell you that the Professor of German Literature was as speechless as I had been flabbergasted.

It is really quite entertaining and enlightening to learn about other departments and how they operate. After all, a good education consists of exposure to different modes of discourse. University-wide committee work is one vehicle for such exposure, and it can broaden you in unexpected ways. In any event, it's part of your duty roster, and you should expect to spend some of your valuable time in this pursuit.

2.14 What Goes On at Faculty Meetings?

A faculty meeting is a gathering of equals. Fellow faculty are getting together to make decisions about the path that the department will take in various matters. It may be noted that some of these equals may be more equal than others. Distinguished Chair Professors may have a louder voice than ordinary citizens. Assistant Professors may not be heeded as carefully as senior members of the department. There may be a secretary at the meeting, but everyone else will be faculty.

Of course the Chair will officiate at the meeting. If the Chair is in fact a Chairperson (see Section 6.1), then he/she serves as a facilitator at the meeting. The Chair in this case is no different from anyone else—except that he/she is standing in front of the room and perhaps he/she is likely to be better informed. The faculty and the Chairperson make decisions together, and the Chairperson implements them. If instead the Chair is a Head (again see Section 6.1 for this distinction), then in actuality the Dean has elevated him/her to a quasi-exalted position. The Head is actually answerable only to the Dean, and can make decisions math majorously. The purpose of the faculty meeting in this case is for the faculty to become informed and to make recommendations to the Head. But the Head makes the final decisions and acts accordingly.

The primary vehicle for decision making at faculty meetings is *discussion*. If the subject at hand is something routine—like what brand of chalk to buy—then the discussion is liable to be brief and peremptory. Likely as not, the Chair will say, "We've looked into the chalk situation, here are the major brands, here are their attributes, and here is what we recommend." Somebody, just to be heard, will say, "Is that the most cost-effective chalk that can be had?" The Chair will say, "Yes," and a vote will be taken. The entire matter can be dispatched in about five minutes.

More exciting is a decision regarding a tenure case, or perhaps a major change in the way that calculus will be taught. Or a decision regarding hiring. Any of these is liable to engender vigorous discussion, and perhaps heated disagreement. People can get quite emotional over these matters, as they will feel that these are decisions that can shape their lives. Those who have weight to throw around may plainly heft it for all to see. An endowed Chair Professor may remind others of his/her status. Some guy/gal who just got a $2 million grant may remind people that (by some yardstick) he/she is now the most visible person in the department. Others may cite their longevity or their well-established record of departmental service.

A good faculty meeting can be exciting and invigorating. In the best of all possible worlds, you will come away from it with a newfound respect for your colleagues, and an uplifting feeling of decisions well-made. Another possibility (usually the exception) is that you will be angry with your colleagues and mad at the world.

Some decisions—the small ones—at a faculty meeting can be made by consensus. Others will take a show of hands. The really delicate ones are handled by secret ballot. If it is a question that will require considerable discussion outside the meeting, then the Chair will instruct the group to get the ballots into his/her secretary by Friday at noon (or another carefully chosen time). Otherwise the ballots will be collected on the spot.

Sometimes there are just topics that require discussion, with no particular decision to be made. In this instance the purpose is to inform people about a matter that affects everyone. Perhaps a change in the college curriculum. Perhaps a change in University admission policies. Perhaps a proposed change (by the Dean) in teaching loads (*that* is liable to engender some spirited commentary!).

So a faculty meeting is also to inform and to focus on questions of the day. Many departments will have a faculty meeting every month or two. The smaller teaching departments tend to have faculty meetings every week, or even more often. I know of some departments where the calculus exams are *written* at faculty meetings. I know of a good Physics Department in which the entire department has lunch together *every day*—and attendance is *expected.* You might be excused for a dentist appointment, but generally you are supposed to be there. And departmental business—as well as physics—is discussed.

Every department will have its own method of governance, its own rules of operation, and its own customs. You will learn them soon in your new working environment, and you will become accustomed to them. They are part of your life, and they are something you will want to be a part of.

2.15 Serving as a Mentor

Of course if you are just a tyro at the job—whether it is academic or industrial or governmental or other—then it is not likely that you will be put into a mentoring role right away. This will come later in your career. Some departments have a formal mechanism for assigning a senior faculty mentor to each new junior faculty person. Many others do not, and your mentoring experiences will come about through ordinary social intercourse.

My own view is that the mathematics profession has been somewhat derelict in providing adequate mentoring for the beginning members of our tribe. If you are faced with an opportunity to mentor, then you should embrace it and do what good you can. You will not get a lot of brownie points for being a good mentor; the reward is in the satisfaction of helping someone along (just as you would have appreciated being helped along). It is worthwhile, and I recommend that you do it.

Good mentoring requires patience, perspective, fairness, and sound judgment. Many people will take a mentoring transaction as an opportunity to criticize and rat out their colleagues. This is tempting, but a dreadfully bad idea. First of all, you should never wash your dirty linen in public. Second, it makes you look cheap and unprofessional to behave in such a manner. Thirdly, you would not want your colleagues talking about you in such a fashion; so show the same courtesy in reverse.

What might a young person ask of you as a mentor? Some sample topics are:

- How do I organize my new class? What preparation (apart from putting together my lectures) need I do? How do I write exams? How do I grade fairly? How do I manage my TAs?

- How do I launch and develop my research career? How do I find good problems to work on? With whom should I try to collaborate?

- Where should I publish my work? How do I choose a journal? What is the procedure for submission of an article? How long should I wait to hear a decision?

- Where should I apply for g rants? How do I apply? Do I do so alone or with others? What agencies will be most receptive to my requests?

- Should I go to a lot of conferences? Is this a good use of my time? Should I try to speak at conferences?

- Is it useful to go to the annual AMS/MAA meeting in January or the annual SIAM meeting in July?

You get the idea. Some of these you can answer right off the top of your head, and others you will have to ponder before you can say anything useful. You want to try to help, as the young person who approached you probably really needs the help. In a sense it is part of your professional duty to do so.

2.16   Undergraduate Advising

Part of your job as a math professor is to be an advisor or counselor to some undergraduates. Usually there are people in the university administration building who are full-time professional counselors, and they will do a certain amount of student advising, especially with freshmen and with students who are having academic difficulties. But somebody has to advise the math majors, and that somebody should have some technical knowledge.

Often advising consists in just reading the university or college catalogue a little more patiently and a little more carefully than the student. Of course your seniority and years of experience at institutions of higher education will play a significant role here. If you are going to be a truly effective advisor, then you need to take a few minutes to acquaint yourself with the breadth requirements, the prerequisites for the major, the different versions of the major that are in place, and what sorts of courses are required for each flavor of the major. If you are a relatively new faculty member, then it might be a good idea for you to meet with the Vice-Chair for Undergraduates, or with some senior and experienced faculty member, to learn the essentials of undergraduate advising at this particular institution.

Of course a bright student can indeed read the catalogue and the course guide just as well as you can. So often the sort of advice that a student will want may be, "Should I take this course or that course?" Or "Should I take this instructor or that instructor?" You are probably well equipped to answer either of those queries, but you should endeavor to be diplomatic with the latter.

Another fairly common type of question is, "The major seems to require this computer science course, but I would like to substitute this other course instead." I never have felt competent to answer such questions, nor did I feel empowered to make such a judgment. So I would refer the student to the Vice-Chair for Undergraduate Studies. This action is appropriate for some other sticky questions as well, especially ones where a policy decision needs to be made.

Generally speaking, a student who is your advisee is *required* to see you at the time, or just before, he/she registers for classes. And then you are required to provide a signature—or perhaps click a box OnLine—so that the registration can go through.

On occasion—and you can only hope that these will be rare occasions—student advisees will bring you their personal problems. These could be troubles with loved ones, or parents, or partners, or roommates. They could be problems with drugs, or problems with the law, or medical difficulties. This is dangerous territory, and I would advise you to refer the student to a professional counselor—there are plenty of them in the Administration building, and this is why they were hired.

One rewarding and entertaining aspect of undergraduate advising occurs when you are asked to direct an honors thesis. Then you are usually dealing with a bright and capable student with many interests. Your job is to help the student pick a good topic, give him/her some things to read, and then answer questions as the student works through the material. Of course you are obliged to read the thesis and make editorial comments, and ultimately you will be required to approve it. The experience is usually satisfying and educational and I recommend that you give it a try.

2.17   Graduate Advising

I discuss Ph.D. dissertation advising elsewhere in this book (Section 4.14), and see also [KRA3]. But there are other types of graduate student advising.

Sometimes students who are struggling with the qualifying exams[13] seek

---

[13]As you no doubt know from your own experience in graduate school, the qualifying exams are one of the important hurdles in basic graduate studies. The student takes a set of exams to demonstrate proficiency with basic areas of mathematics. More is said on

advice. Generally speaking, it is the job of the Vice-Chair for Graduate Studies to handle such matters, but you may be called upon because of your area of expertise. Sometimes you can give some tips on what to study, or how to study, or how to manage one's time.

I have had graduate students—with whom I was friendly for other reasons— seek out advice because they were having personal or financial difficulties and were afraid they were going to have to drop out of the program. I was happy to be able to provide a little guidance, and to offer some suggestions and alternatives.

I have mentioned elsewhere in this tract that my own university has a system of two oral presentations that each graduate student must perform— this occurs after the quals and before the thesis. The student will have a faculty mentor for each presentation. This is a great opportunity to get to know a student who might become your Ph.D. student, and it is usually a pleasant experience.

Generally speaking your position as a professor makes you a sort of cultural icon and you could be asked advice by almost anyone about almost anything. I have been phoned up and asked questions (by the Electoral Office) about designing ballots and counting votes. I have been asked questions (by lawyers) about the probability of one hive of bees stinging a victim rather than some other hive of bees stinging that same victim. I have been asked advice about making machine dies of a particular geometric shape. I have been asked advice by the Chief of Campus Police about how to put together the statistics for a report he was writing. I have been asked by the geography department how to calculate areas on a large map. And students have asked me almost any question you can think of about almost any topic you can think of. It is one of the more recreational parts of the job.

2.18   Your Role in the Professional Societies

I have a rather stodgy view in this matter. The professional societies— the American Mathematical Society (AMS), the Mathematical Association of America (MAA), the Society for Industrial and Applied Mathematics (SIAM), the Association for Women in Mathematics (AWM), the National Association of Mathematicians (NAM), the Society for Advancement of Chicanos and Native Americans in Science (SACNAS), and many others—are

---

this topic in the reference [KRA3].

essential to what we do, and they help us to do it. The AMS, for example, holds many important conferences (the January joint meeting with the MAA is perhaps the biggest), publishes many fine book series, publishes important scholarly journals, sponsors and develops `MathSciNet`, and develops many professional tools to help people to apply for jobs and to further their professional lives. The MAA concentrates on teaching and dissemination issues. SIAM is the organization for people who care about applications of mathematics. SACNAS endeavors to meet the needs of mathematicians who are members of underrepresented societal groups. NAM concentrates on professional needs of African Americans.

I feel strongly that you, a budding mathematician, should belong to one or more of these professional societies. Membership is relatively inexpensive—about $160 per annum as of 2008 (in fact AMS dues are reduced for the first few years), or the cost of taking your spouse out for a nice dinner. And the benefits are considerable. Plus it is important that you be part of the infrastructure of the profession. Also, you will receive in the mail (and by e-mail) useful periodicals that will help you to keep up with professional developments. Finally, the professional societies can serve as a touchstone for helping you to network with your peers.

It is an observed fact that young mathematicians these days have a decreased interest in belonging to the professional societies. This is too bad; it is their loss, as well as ours. It is the upcoming generation that will keep the societies going. The AMS and the MAA are each about one hundred years old (SIAM and AWM and SACNAS and NAM are considerably younger), and it is no accident that American mathematics rose to considerable world prominence during those one hundred years. The professional societies played a notable role in the process, and they will continue to do so as we chart our course in the twenty-first century.

2.19 Translators

One important service that a mathematician can provide to the profession is as a translator. It has come about, even during my time as an academic, that English has become the predominant language in the subject. Even the French—who tend to be quite chauvinistic about their lovely language—tend to publish their articles in English these days. It has been a tradition in good graduate programs to require all Ph.D. candidates to show proficiency in two

languages—usually chosen from among French, German, and Russian. But
even that venerable habitude is falling by the wayside.

There are automatic OnLine translators—`Google` has one—that one can
use for free to "translate" from almost any language to almost any other.
These tools do not work very well, and they are particularly poor with technical languages like mathematics. You cannot depend on a computer translator
to help you read a math article.

Certainly there are a number of journals these days—journals which are
basically "English" journals—that will accept papers in French or German.
There is one journal that will accept papers in Latin, and another that will
take papers in Esperanto. The fact remains, however, that English is now
the norm.

The greatest demand for translating in our discipline—at least in modern
times—has been for Russian and Chinese articles. But in the late 1980s a
good many Russians left the former Soviet Union and moved to the West.
Many Chinese left China. Most of these scientists now publish in English.

And in fact the dissolution of the Soviet Union caused a considerable
realignment of the Russian mathematics journals. Most of these journals were
moved to the London Mathematical Society for translation (this decision was
made largely for economic reasons). Of course these days many companies—
especially publishing companies—outsource their work all over the world. So
it is probable that a good deal of translation from Russian into English is
done *by Russian scholars in Russia*. Whereas the AMS used to translate all
the principal Russian mathematics journals into English, now it does just
three of them.

So why is there any need for translation of mathematics? There are
actually several:

- There are still articles in Russian and Japanese and other exotic languages that one wants to read.

- If you publish a good book—one that students will want to read—in
  your native tongue, then students in other countries will probably want
  to read it too. And, though they learn English in school, they are really
  not all that comfortable with it. They would like a text in *their* native
  language.[14]

----

[14]I still recall when my graduate algebra professor adopted Bourbaki's book—*in French,
of course*—for our text. I was so exasperated that I refused to read it. This was probably

- Publishers are in the business of selling books, and arranging for a translation is often in their best interest.

The long and the short of it, then, is that publishers will need translators. These will include book publishers, journal publishers, and professional societies. They will pay people to do the translating. Usually they want a Ph.D. mathematician for the job, because they don't just want the words translated—they want the *sense* translated. Mathematics is subtle. Just for example, the use of "not" and of negative statements is quite delicate and must be handled properly. Definitions must be just right. Proofs must cohere. So a properly trained person must do the translating.

The trouble is that the budget for publishing advanced math books is slim. A popular text might sell 200,000 copies, and the publisher's profit on each unit (after the initial 2,000, which pays for the fixed costs) is about $20, so there is plenty of cash around for translating or anything else the publisher might want to do. But an advanced math book might sell only 1,000 or fewer copies. After the fixed costs (typesetting, overhead, etc.) are covered, the profits are slim. So publishers think of translation in that context as really a luxury. And what they can pay for this service borders on paltry. A publisher will offer a few thousand dollars to translate a book. Translators of journal articles can make $15 to $25 per page, depending on who is paying and on the precise nature of the work. Some publishers pay by the line in the target language; they pay literary translators by the word. That can be a tidy sum, but these people are in the minority for sure. If you are the translator, and if you convert the typical remuneration into dollars per hour, you will not end up dancing in the street. Compared to what your attorney or your physician is paid, this is pretty slim pickings. If you do translation, you are doing it mainly as a service to the profession.

2.20   Is Mathematics Just a Service Department?

It is a sad but true fact that many administrators, when they think of the math department, have as their first thought that, "Oh, yeah. Those are the guys who teach calculus." You can bet that their primary consideration is *not*, "Oh yeah, those are the guys who proved the Bieberbach Conjecture."

the worst math course that I ever experienced (and that was clearly my own fault).

Mathematics plays a central and vital role in undergraduate education on campus. We, as the key players in this operation, should be well aware of this fact and capitalize on it. Typically, mathematics teaches more undergraduate student hours than any other department. Also we teach more students who are there under duress than any other department. Students in freshman chemistry may hate the bloody course, but they certainly know why they have to learn chemistry (often they are chem majors or pre-meds or chemical engineering majors). Students in an education course are almost all education majors, so they certainly do not question their curriculum. Students in eighteenth-century French literature are probably there by choice. In many majors, math is a requirement just to make the curriculum more rigorous; it serves no practical purpose in the subject matter—at least the way it is presented in a modern undergraduate curriculum in this country. The mathematics department teaches a great many students, and it teaches a great many students who would certainly rather be doing something else. And who do not really understand, in a visceral sense, *why* they are being required to take mathematics. This is a difficult audience to face, but facing it is what we must do as mathematics professionals.

The upshot of the message in the preceding paragraph is that you, as a mathematics instructor, have a special obligation to make your lectures/classes sparkle, to make the material interesting and relevant, and to relate the mathematical ideas to ideas in other disciplines. Find a natural way to weave into your message the idea that mathematics is a part of life, that it is useful, and that it is exciting. Make it speak—indeed sing—to your students.

In addition to its key role in the curriculum, mathematics is also a critical part of modern technology. Most aspects of computer construction and design are mathematical. Ditto for the main ideas in cryptography, in medical imaging, and in many other hot fields. Calculus is a vital part of physics, economics, and population dynamics. Statistics is central to social science, engineering, and physics. Probability is the language of quantum mechanics, the social sciences, and finance. Discrete mathematics plays a vital role in information technology, biology, counting theory, and many other fields. Mathematicians make an essential contribution to the quality of life in this country. We are scholars of considerable gravitas.

So, yes, we *are* a service department. We give great service to the college, and to the university. But we are much more. We are at the forefront of modern scholarly thought. We are serious researchers engaged in important investigations. We have research grants, we receive invitations from all over the world, we are widely published, and we give considerable service to the mathematics and science professions. We are recognized academicians with worldwide reputations. We are an important part of the professoriate, ones who make notable contributions in many dimensions of the discipline.

Math is part of the traditional core curriculum at a college or university. It is one of the standard "three Rs". It is one of the liberal arts of a classical education. Most people arrive at college or the university having already seen quite a lot of math in their lives. Sad to say, many have already decided that they really don't like math. This is the audience that we face.

When the math department Chair meets with the Dean to discuss teaching issues, he/she may try to convey some of the information in the preceding several paragraphs. And he/she may or may not succeed. Deans do not like to hear (what they perceive as) excuses. What the Dean wants is *results*. If the Dean holds in his/her hand statistics that seem to show that math department teaching evaluations average distinctly lower than English department teaching evaluations, then he/she is going to want to know why. And he/she will *not* be receptive to an answer like, "English is easy but math is hard." or "The French department grades easier than the math department." or "The students in history like history and want to be taking the course. The students in math are there mostly to fill a requirement for their majors."

It is certainly true that the typical professional mathematician has received little if any training to be a teacher. I—and many others—consider myself to be a model teacher, but I spend precious little time instructing my Ph.D. students on how to teach. I guess, truth be told, I depend on the math department to worry about this matter. And, thankfully, my department does so. We have a *mandatory*, three-credit-hour, semester-long course on teaching that every first-year student takes. And I think that it does a lot of good. No graduate student TA at Washington University is allowed to stand in front of a class until he/she has taken that course in teaching methodology. When I was Chair, I taught the course myself—in part to see that it was done right, and in part to send the signal to the students and to the entire department that *we take teaching seriously around here.*

There are several people in the Washington University mathematics department who are teaching heroes, and are well-known around campus as such. That is a real plus for the department, and helps to assuage the negatives that were adumbrated above. Again, it is the job of the Chair to explain the department to the Dean. He/she must help the Dean to understand that we are a group of committed scholars who take our teaching duties seriously and dispatch them with professionalism and some flair. We have strengths in research, strengths in teaching, and strengths in department and university service. We frequently communicate with and work with the engineers, the physicists, and other units on campus. We have broad-ranging interests, and we make many types of contributions.

As department Chair, I was once at a meeting—that the Dean called, without telling me what the topic of discussion was going to be—in which the Dean began (with raised voice) by telling me that the math department's priorities are

1. research

2. the graduate students

3. the math majors

4. students who might become math majors

5. everyone else

The Dean then fixed me with a baleful stare and declared: "I want you to reverse the order of these priorities!"

I concluded at that moment that the Dean was a blithering idiot. But I was confronted with a colossal misinterpretation of our department's dynamic and value system by someone in power who should have known better. I had to deal with it. I had to explain that we teach more undergraduate student hours than any other department, we are among the top ten departments for the number of undergraduate majors that we have, we have many innovative features in our curriculum, and we have many committed teachers. Many of us have to conduct our research programs almost as a hobby because we have so many other duties—mostly connected with teaching—in the department. One of the main duties of the math department is try to help students who are *not* math majors, who do not have any particular flair for mathematics, to come to grips with the subject.

I frankly do not know whether I convinced the Dean of anything, or whether I converted him to the true path.[15] But I can say that he backed

---

[15]The fact of the matter is that *we* were already on *his* true path, but he had to be made privy to this fact.

off and never got in my face about this set of issues again. I never found out whether he delivered a similar harangue to any other science Chair. Given the way that universities and their administrations work, I would doubt it.

The point is that this is the world we live in. The Dean is, likely as not, a lab scientist (lab scientists by nature must have people and organizational skills) and certainly appreciates the value, and the contributions of, the Chemistry Department and the Biology Department. Not so with mathematics. For us it is an uphill battle to establish our worth and to maintain people's perception of it. Even if the Dean pretends to understand where we are coming from on day n, he/she will have forgotten by day $n+5$. And will have to be convinced anew. Such is the lot of the math department Chair.

Chapter 3

Sticky Wickets

When you become famous, being famous becomes your profession.

James Carville (political consultant)

I'm lucky to be in a profession where you can keep getting better.

Cy Coleman (songwriter)

You're in a profession in which absolutely everybody is telling you their opinion, which is different. That's one of the reasons George Lucas never directed again.

Francis Ford Coppola (movie director)

I love writing and do not know why it is considered such a difficult, agonizing profession.

Caroline B. Cooney (author)

As human beings, we are endowed with freedom of choice, and we cannot shuffle off our responsibility upon the shoulders of God or nature. We must shoulder it ourselves. It is our responsibility.

Arnold J. Toynbee (historian)

Politics is a profession; a serious, complicated and, in its true sense, a noble one.

Dwight D. Eisenhower (president of the United States)

Not everything that can be counted counts, and not everything that counts can be counted.

Albert Einstein (physicist)

I'm all in favor of keeping dangerous weapons out of the hands of fools. Let's start with typewriters.

Frank Lloyd Wright (architect)

3.1   How to Deal with a Sequence of One-Year Jobs

When the job market gets tough, it is usually because of economic strictures on universities.[1] They do not have as many two- and three-year postdocs and (potentially) long-term Assistant Professorships to offer as we would all like. But the schools still need to staff their courses. An unfortunate upshot of this situation is that a variety of last-minute, one-year, non-renewable positions get cobbled together. You may be the unfortunate recipient of one of these.

I know people who began their careers with a sequence of one-year positions like this. They really suffered. As I have said elsewhere in this book, you cannot do mathematics effectively by grabbing snatches of time here and there. You really must have large chunks of time in which to concentrate on your work. And these chunks of time cannot occur once per month. They have to be a regular part of your life. You must also have some stability and some peace in your life. Bouncing around from job to job is not conducive to such serenity.

When you have one-year positions you are always packing your bag or unpacking your bag. You are constantly going through orientation procedures, learning new work schedules and new computer labs and new teaching paradigms. Your life is fairly chaotic. You are constantly engaged in the duress and indignity of applying for new jobs. It is not like a one-year sabbatical, which is designed to help you relax and learn new things. It is more like a one-year stint in hell.

There is no easy answer to this problem. If you need a job, and you want to stay in academics, then this may be all you can get for a while. I will say this (as an alternative point of view): Taking a job in the private sector, or working for government, is not a one-way door. Just because you go to work for Microsoft or Hughes Aircraft for a while does *not* mean that you cannot return to academe. I certainly know people who have done it. And you may find that you are happier (and often better paid), with more peace of mind, taking another kind of job for a while. You will have a little more time to

---

[1]There are other factors that can play a role. For purely political reasons—Tiananmen Square and the collapse of the Soviet Union, for example—there was a great influx of Chinese and Russians and Eastern Europeans into the American job market in the late 1980s. Many of these immigrants were world-class scholars whom everybody wanted. These events had a profound impact on the availability of jobs for Americans.

get your life organized and to plan your future.

In particular, you should focus on managing your time and your career development. Get some research done—stuff that you can put on your CV. If you can, avoid committees and other time drains. Gear your life to looking your best for your next job search sortie.

3.2 What to Do If You Cannot Get Along with a Colleague

If you are an Assistant Professor in a math department (or perhaps a junior colleague at an industrial firm) and you cannot get along with your mentor or supervisor or Chair, then you have a problem.

First, you should examine your conscience and determine whether this lack of collegiality is at least partly your fault. If it is, then do something to fix the situation. If it is *not* your fault, then you have some choices. In the industrial setting, the mentor is probably assigned to you (i.e., he/she is likely your boss) and you are stuck with the person. In the academic setting, you may be able to find another mentor. However, if you can't get along with your Chair then look out. The Chair is going to be assembling and promoting your tenure case. The Chair decides your raises and your teaching assignments. In short, the Chair is the nearest thing to a five-hundred-pound gorilla that you will see for several years. Make it your mission to have a good relationship with your Chair.

If you are really stuck with a boss in the private sector who is difficult or impossible to work with, then try to stay away from him or her. Develop good working relationships with others. You may be able to make a lateral move (*not* initiated by you) and get out of the situation. Otherwise you are going to have to swallow your pride and live with it for a while.

More generally, it can certainly happen that you will have disagreements with colleagues. This is what serious discourse is about: If everyone agreed with everybody else about everything, then life would be pretty dull. Ideally, the folks in the math department should be able to have vigorous discussions, disagree about interpretations of facts or matters of policy, come to some decision, and then go off and have lunch together. Unfortunately, this does not always happen.

I must return again to a recurring theme in this book. Whereas most of us have little control over a number of the important parameters in our lives—taxes, our home lives, international politics, the pursuit of a cure for cancer—we mostly tend to feel that we have a *lot* of control over what goes on in the math department. This is our little self-created world. We get to decide the curriculum, we get to decide whom to hire, and we get to decide whom to tenure. If anyone—including a colleague—dares to encroach on this purview, then there is hell to pay. Just as an instance, there was a huge brouhaha over a tenure case in a major research mathematics department about twenty years ago. It got played out in the newspapers, and there were lots of lawyers and lots of suits and countersuits. It became apparent over time that some key administrators had made some dreadful errors in handling the tenure case, and things got quite ugly. In the end—after a protracted battle of over seven years—the Chancellor intervened and granted the candidate tenure and back pay and damages. I can tell you that the department was vehemently split over this tenure issue. People felt that they were not simply differing over matters of fact, but instead over much more fundamental questions. The entire affair was a life-changing experience for many of the faculty. And today, even twenty years later, many of the wounds are still raw.

Another issue that has split math departments in the past twenty years is the question of teaching reform and calculus reform. Recall that, in 1986, the inimitable Ron Douglas organized a small meeting (about twenty-five people) at Tulane University in which he asserted that calculus teaching in this country was a disaster, the attrition rate was dreadful, and the failure and dropout rates were a matter of national shame. He provided compelling statistics to support his contentions. And he uttered the fateful words, "The lecture is dead." Thus was launched the so-called "Reform" movement in mathematics teaching. And departments were quickly split. Many of us old codgers were not in the mood to hear that a teaching method that we had comfortably used for thirty or forty years—and which in fact has been used in many different cultural contexts with many different audiences for over 2000 years—is now to be thrown out the window and replaced with *what*? Well, what was proposed was (among other things) **(i)** group learning, **(ii)** self-discovery, and **(iii)** computer-aided learning. Several new texts, most notably the Harvard Calculus Project, were produced in support of the new movement. There were also notable Reform projects at Duke, at the five colleges in Massachusetts,[2] and in Oregon. Other projects too numerous to mention proliferated all over the country. And there was copious government money to subsidize the changes that were taking place. Again it happened—and I describe this phenomenon elsewhere in the book—that people who had been fairly quiet in the mathematical life up until now suddenly had $3-million-dollar grants. That is quite a lot of money—enough to set up an empire in which the project would have its own offices, its own staff, its own students, and its own agendas.

Just as people sharing drinks in a bar will go to the mat over whether the Yankees or the Mets are the better team this year, so people just got hysterical over Traditional versus Reform teaching. The scholarly *Chronicle of Higher Education* reported that math faculty were physically assaulting each other over the question. People who had been working away for many years to prove their theorems and occasionally get paltry grants from the NSF were resentful of these upstarts getting multimillion dollar grants and exerting so much influence. And of course the Dean and the administration don't know much about mathematics and math teaching. But they know a lot about money. When they gaze from a distance at the mysterious math department—full of enigmatic eggheads who do something or other in some extremely recondite realm of human thought—the guy/gal with the $3-million-dollar grant really stands out.

Of course fair is fair. These folks with the big grants will also play a big role in the teaching enterprise of the department—and give it a high profile. These people will look after calculus and precalculus and other knotty courses. They will tend to the mentoring of freshmen. They will design courses and write syllabi. In short, they can make a substantial contribution and make everyone's life easier.

I myself think that it's fine to disagree with people. This is a healthy aspect of being a well-educated, intelligent person; you discuss and you learn. If it's done right, everyone comes away with a new and heartfelt respect for each other. But things can go awry. Feelings can be profoundly hurt. People can feel that they have been cut to the quick in a most visceral sense. They can decide to no longer speak to each other. And that is really a shame. The math department is your extended family. You've got to live and work with

---

[2]These are Amherst, the University of Massachusetts, Hampshire College, Mt. Holyoke, and Smith. They are all within shouting distance of each other in western Massachusetts. In fact there is a bus service that will take you from one campus to another.

these people for forty years or more. Best if you can all get along. When
things are not going well with your biological family, there are professionals
(priests, rabbis, counselors, psychologists) who can help. In the academic
setting it's not clear where you are supposed to turn for succor. One option
is the Dean, but likely as not he/she is part of the problem.

I once had a huge fight with a colleague over a matter of departmental
policy—something to do with hiring, I think. In fact it had to do with
applied math, which has been another hotbed for controversy over the past
forty years. Don't get me started. This guy and I really liked each other a
lot, but we disagreed in fundamental ways about this particular issue. And
we could be found in the hall yelling at each other at some length. It really
got ugly.

Fortunately, the goodness lying at the bottom of our souls finally got
the better of us. More significantly, being the mature individuals that we
obviously were, we both realized that this bickering over a secondary matter
was interfering with the collegial and effective running of the department.
Neither of us was enjoying the backbiting and squabbling. At some point we
decided to kiss and make up. We both bent over backwards to repair the
damage and set our relationship back on an even keel. In fact, we started
working on a problem together (we were actually in two very different fields),
and we ended up writing two good papers—one with Paul Erdős! So there
are some happy endings, and we could use more of them.

3.3   What to Do If You Cannot Get Along with Your Students

If you cannot get along with your undergraduate students, then it probably
is your fault. Most undergraduates are delightful, are happy to be in college,
and are willing to make at least a token effort to play the learning game.
They are generally quite easy to get along with. If, on the other hand, you
are one of those people who feels compelled to come across as a hard-nosed
autocrat, then you may find that you have few friends.[3]

---

[3]One of my colleagues habitually begins a large lecture class—first thing on the first
day—by laying down the law. He enunciates that teaching a large class entails particular
managerial problems, so he has special rules. Among these are: **(a)** You cannot come
to class late; **(b)** You cannot leave class early; **(c)** You cannot hand in homework late;
(d) You cannot eat in class. And on and on. Imagine what kind of tone this sets for the

It is true—very rarely—that you can get a class that is just a dud. Somehow the chemistry is lacking, and you all hate being in the same room together. You can take strength in the fact that the semester is only fifteen weeks long (if you are on the quarter system, make that ten weeks!) so—unlike a bad marriage—it will soon be over. But you can also try varying your teaching style or your lesson plan. Use some guest instructors—*not* to torment them, too, but to introduce some variety into the learning situation. Show a film. Have a discussion. It's like trying to carry on a conversation and finding that the situation is cold and lifeless. There are a variety of devices that you can use to attempt to bring it back to life.

Students are different from you and me. They are much younger, and their values and goals are quite different from the things that you and I worry about. Their expectations are in many instances unfamiliar to us. But we were there not so long ago. With a little perspective, we can learn a little tolerance and understanding of the student point of view. And it can be refreshing, and provide a new look at the world around us. The main point is to be open to what the students have to offer; and to respond in a constructive fashion.

Teaching—good and effective teaching—is of necessity interactive. Were this not the case, then we could all be replaced by videotapes or OnLine learning. If something is going awry in your class, then *work with the class* to fix it. You should have a good, working relationship with at least some of the students. You should be able to talk to these people, take the temperature of the situation, try to dope out where things are off-kilter and what you can do to fix them. Mid-term, anonymous teaching evaluations can be helpful. But nothing beats talking to people.

3.4 What to Do with a Problem Graduate Student

Usually you will not accept a student to study for the Ph.D. under your direction unless you know that student fairly well. After all, this is a long-term commitment, and an intense relationship. It is going to take a good deal of your time and effort. So this is not an obligation that you are going

---

semester. Imagine what kind of impression this makes on the students. I think that you can do better.

to take on lightly.

You can get to know graduate students through classes, of course. Many graduate students will sign up for a reading course with a professor whom they might be thinking of asking to direct the thesis. At my own university, we have a system of oral presentations that the student gives *after* the quals and *before* the thesis. The student has a mentor for each of these activities, and frequently the mentor for one of them ends up being the thesis advisor.

It really is not wise to take on a Ph.D. student cold. You need some mechanism to find out about the student's work habits and personality. You need to see whether the two of you can communicate, and work productively together. You are liable to spend five years in very close contact, working toward a very large and formidable grail. Make sure that this is a partnership that is going to work.

If you do this right, you probably go into the thesis-advisor relationship with a good idea a priori whether you will be able to get on with this graduate student in the long term. Nonetheless, as with all aspects of human relationships, things can go wrong. I have had sixteen Ph.D. students, so can count myself among the more successful thesis advisors. But I did have one graduate student who started leveling death threats against me—in posters that he hung all around the department! The institution can provide some help in a problem situation like this one. There are campus psychiatrists and counselors who have considerable experience dealing with the peculiar stresses and disorders that come from academic life. It is easy to have a session with such a person, and he/she can make a number of useful suggestions.

Most likely your graduate students are not going to be taking out contracts on you. But they can become blocked and hide from you (which is just the opposite of what they *should* be doing). Some students are innately lazy—or have never been challenged to really work before—and they just dig themselves into a hole. Graduate students can also develop substantial personal difficulties—like having an intimate relationship with a student in one of the classes that they TA. You, as a faculty member, are not necessarily qualified to deal hands-on with all these situations. But if you have a good working relationship with your student, then you can get the student pointed in the right direction—towards counseling, or self-discipline, or whatever the situation may call for.

Your more experienced colleagues, who may have had similar trials themselves, can probably give you some tips. A student who is mentally unstable and perhaps disruptive is probably not doing well in his/her studies either. It might be appropriate to get him/her a terminal Master's Degree and gently ease him/her out the door. If appropriate, steer the student toward professional help. I have dealt with university counselors and psychologists, and they are remarkably shrewd and resourceful people. They have clever and persuasive means of meeting troubled students "by chance", engaging them in conversation, and getting them some help. Perhaps the department has some experience with difficult cases, and can give you some guidance.

The main point is that you should *not* try to deal with a difficult situation like this in isolation. Talk to people. Get some guidance and direction.

3.5   Jobs in Industry

I have seen good books, well-meaning books (such as [BEC]), refer condescendingly to the "transition to industry." The unspoken implication is that you couldn't make it in the academic world so had to fold your tent and take a job with Microsoft or Daniel Wagner Associates.

I am happy to say that in today's world this is not the right attitude, and not the right way to look at things. It's been a long process, but we in the mathematics profession now realize and acknowledge that a mathematics student today faces many exciting opportunities and many choices. The tired old phrase "If you study math, then all you can do is teach" has never been further from the truth.

A Master's or Ph.D. student today certainly can choose the academic path, but he/she could also become an actuary, could work in the genome project, could work in aspects of law, could work for the National Security Agency (NSA), could work for Hughes Aircraft, or in the financial sector, or many other choices. Three of my former Ph.D. students work at the National Security Agency NSA. One is an Internet venture capitalist (and he used to be a tenured Associate Professor at Tennessee Tech), and one is in software development.

The truth is that—if one wished to take a rather dyspeptic point of view—academics is perhaps not quite so attractive these days as it once was. It is far too competitive and really not so friendly. Salaries are little more than adequate, and opportunities for advancement are limited. Insidious ideas from the business world like Total Quality Management (TQM) are invading the way that we run our schools, and university administrators are becoming less like academics and more like drill sergeants. Colleagues can be brusque and unpleasant and working hours are long. Students can be rude, tedious, lazy, and uncooperative. Deans and Provosts (and even department Chairs!) can be a bore. The ordeal of trying to get tenure and then battling for promotion to full Professor can put a considerable strain on your marriage and your personal psyche. It just may not be for you. Working in the private/industrial sector may prove to be much more attractive.

In academics you have two big promotions in your life—from Assistant to Associate Professor and from Associate to Full Professor. Each of these is accompanied (usually) by a notable raise. Apart from those two life-changing events, your raises in the academic life will be modest (absent getting an outside offer or winning a big prize).[4] Life is different in industry. There you will probably get a raise every year, and you may be up for promotion every four or five years. The rewards may be considerable, especially once you start moving up the management ladder.[5]

In industry you typically don't have complete discretion in the problems that you work on. You may have a *choice*, but the choice will be among problems that matter to the company—that their funding sources are willing to pay for. You are no longer thinking and researching for the joy of the game. The point is to get results. It is a different sort of process, and it has its own rewards. One of my former Ph.D. students who works at NSA likes the work, and finds it satisfying, because she feels that people *actually use* the things that she creates.

People experienced with working in the private sector (see [BEC]) point out that academic recruiters are looking for "smart" while industrial recruiters are looking for "team players who can make a contribution." This is a bit simplistic. No recruiter is looking for "dumb." And everyone wants to hire a team player. Especially if you are applying for a job at a school where teaching is a major part of life, your academic qualifications (your research specialty, the particular school where you got your Ph.D., the bona fides of your thesis advisor) may be of only secondary importance. They will be examining you as a *person*. Often industrial recruiters are looking for

---

[4]When I worked at UCLA, raises did *not* occur every year—not even cost-of-living raises. What would happen instead was that, every three or four years, there was a "catch-up" raise. Most of my fellow faculty could document how their earning power had fallen steadily during their time at U.C.

[5]It is quite common in the industrial setting—although certainly not mandatory— for someone who is talented in R&D (research and development) to be fast-tracked into management. This is where all the power, and the big bucks, is.

problem-solving skills as well as people skills.

Fields important to a non-academic employer may include

- computer science (programming skills)

- financial modeling

- statistics

- operations research

- numerical analysis

- engineering

This does *not* mean that your major subject area needs to be chosen from this list. What it *does* mean is that it helps if you can show that you have had some exposure to some of these ideas and techniques. I myself have never had a course in computer science in my life. But I have created two major pieces of software—software that actually changed the way that certain industries do their business. Were I applying for an industrial job, that would count. I am no expert in operations research, but I have taught a course in linear programming and I know something about Karmarkar's algorithm. That would count. I have collaborated with engineers,[6] and that would be meaningful to an industrial recruiter.

Whenever you are interviewing for a job, you should be prepared to play up your strong points and acquaint the interviewer with who you are and what you have to offer. The interviewer is not going to fish for your best qualities. It is your job to put them forth. Make a case that your skill set meshes well with this company (or school), its people, and its goals. This may be particularly true for an industrial interview, if only because you are trying to create a marriage between your academic background and the more practical working environment into which you propose to move.

3.6 What Do People in Industry and Government Do?

Good question. Much of this work is classified, and I can't describe it (nor do I know it). But I can say a few words about it.

---

[6]The very fact that I can *communicate* with engineers would be noticeable.

Many private companies that do mathematics, or are run by mathematicians (Daniel Wagner Associates is an outstanding example and Mitre Corporation is another), survive largely on government contracts. They work on particular defense projects such as the stealth program,[7] or robot-controlled tanks and jets, or various questions in optimal control and systems science. Those who work for government agencies might work for the Treasury, the Federal Reserve, the Social Security Administration, the National Aeronautics and Space Administration (NASA), or the Census Bureau. Clearly the jobs there involve a lot of statistics and data management. There is a great deal of statistical work in the applied world; data mining is a new set of ideas with considerable interest. A lot of computer graphics. A lot of expert systems and robotics.

The National Security Agency specializes in (but is not exclusively limited to) cryptography. The Institute for Defense Analyses studies a variety of defense mechanisms, including even aspects of psychological warfare.

If you work in the computer industry, then it is not difficult to imagine what you might be doing. I have one friend who worked for Microsoft in developing Microsoft `Word`.®Another friend worked for Sun Microsystems; he wrote a version of their `UNIX` operating system.

Quite a number of mathematicians, or at least people from the mathematical sciences, worked at Xerox Research PARC.[8] This was one of the great cauldrons of creativity in early Silicon Valley days. Founded and run by George Pake of Washington University Physics, Research PARC saw the following innovations in the span of just a few years:

- invention of the computer mouse;

- invention of the laser printer;

- invention of the first WYSIWYG word processor;

- invention of the personal computer (really!);

- invention of the Graphics User Interface (now known as `Windows`);®

- invention of the software now known as Adobe `Postscript`;®

---

[7]This is a program to develop, among other things, jet fighter planes that can be "invisible" to radar.

[8]This is the Palo Alto Research Center.

- invention of the software now known as Adobe `Acrobat`.®

And that is only a partial list. The book [HIL] gives a detailed and exciting history of this facility.

Some companies and agencies have the look and feel of academic departments. For instance, at the National Security Agency one finds that he/she works rather math majorously. Certainly not closely supervised. One has particular projects to work on and one does so, reporting on results only every several months. There are seminars and head sessions much as in a university math department. But there are also differences. At NSA one is not allowed to bring work home (for security reasons). One cannot bring a cell phone to work. One cannot take public transportation to work. When visitors are present, then one cannot talk freely. It is an atmosphere that requires some acclimatization. One can imagine that any company or facility that deals with high-security material will have similar features.

When I lived in State College, Pennsylvania, H. R. B. Singer was a big employer. A number of mathematicians worked there. Singer dealt primarily with electronic warfare, and you can imagine that security was high, and taken *very seriously*. This really is quite different from academics, where everything is wide open. There are plenty of opportunities for mathematicians in the information technology (IT) sector, in medicine, and in finance (to give just three examples). Each of these will have differing levels of security and secrecy.

If you work at a place like NSA, you will probably still write math papers. But much of your work cannot be published in the usual venues.[9] Anything that you want to publish in the "outside world" has to be cleared by a superior. This, too, takes some getting used to. But I know lots of people who work at NSA—the largest employer of Ph.D. mathematicians in the world. They seem to find the work satisfying, and you rarely hear of people leaving NSA.

3.7 What About Tenure?

In academic life, tenure is certainly the holy grail. Typically a career trajectory goes like this:

---

[9]There are special journals that circulate only to people with high-level security clearances in the defense industry. But this is not quite the same as publishing in the *Transactions of the American Mathematical Society.*

- Graduate school (four to seven years)

- Postdoctoral position (two or three years)

- Assistant Professorship (four to six years)

- Tenure decision

- Associate Professorship (four to ten years)

- Promotion to full Professor

- Professorship (twenty to forty years)

The fourth item is the make-or-break step, because if you don't get tenure then you are out of the loop and you will change career paths.

We shall say a good deal more about tenure as the book develops. Tenure is the University's lifetime commitment to you: to pay you a salary, provide an office, create a supportive work environment, provide students, and so forth. There are few jobs in this world that are more secure or more pleasant than a tenured professorship at a university. If the institution *does* grant you tenure, then it is stuck with you for thirty or forty years, and it is in for a financial commitment of $5 million or more. So naturally a good deal of thought will go into the decision.[10]

There is in fact a highly structured, multilayered process for evaluating tenure cases. We describe it in detail in the ensuing sections. The department, the Chair, the Dean, the Provost, the Chancellor, and the Board of Trustees all pass judgment on every tenure case. And this is generally not a decision that is made by your institution in isolation. In most tenure cases the department or the Dean will write out for letters about you—to distinguished faculty at esteemed institutions. They want to know how you fit into the big picture, what is your stature in the profession, what scholarly contributions you have made.

3.8   Sex and the Single Mathematician

We are all sexual creatures, and we are all subject to temptations of the flesh. Many students, many staff members, and many fellow faculty are attractive

---

[10]Of course there are important academic considerations that go into the tenure decision as well. I am taking a somewhat draconian view right here.

and charming and delightful. One may be tempted to get involved with them beyond the ordinary parameters of social discourse. Worse, one may be inexorably smitten and lose all control of one's bodily functions. What to do?

Generally speaking, it is a good idea to keep your personal life separate from your professional life. Not doing so can cloud your judgment, compromise your integrity, and get you into a whole passel of trouble. Never, ever do anything with a student in one of your classes except teach him/her mathematics. Don't go to a bar with your students, don't go to parties with your students,[11] don't play touch football with your students. Universities have *very* strict rules about sexual harassment and breaching those strictures is an entree into a world of grief and pain.

If you are married, then it is certainly not my business to tell you how to live your life. You are supposed to already know the answer. If you are single, then you may very well find that meeting people in the math department is considerably more constructive and convivial than meeting people in a bar or at a dog fight. It may be OK to ask the departmental receptionist for a date, but the date should be completely separate from your relationship in the department. When at the office, you should behave in a businesslike and dispassionate fashion. Likewise for colleagues. Dating students opens up too many possibilities of conflict of interest and (because you have all the power and they don't) harassment, and you would be wise to steer clear of this temptation.

Sexual harassment situations usually occur in private—there are no witnesses. Any adjudication of a sexual harassment suit ends up being a he-said, she-said situation. It can be quite unpleasant and disorienting. It is an observed fact that an awful lot of people—young and old alike, and of both sexes—are simply unaware that their behavior is offensive or inappropriate. In the best of all possible situations, someone will point out to these folks that they need to clean up their act. A number of apologies ensue, and then life goes on as usual. But sometimes, unfortunately, things go too far and lawsuits result. The resulting ordeal can be protracted and can derail the lives of a good many people.

There are certain precautions that one can take to avoid situations that

---

[11]Of course many math departments have colloquium parties, and you should go to them. Have fun. But be aware of where the line is and be aware of the consequences if you cross it.

can lead to misunderstandings related to sexual harassment. Some of these are:

- Never meet with students in your office with the door closed.

- Do not have sofas or other furniture on which one could recline in your office.[12]

- Do not sit close to your students.

- Do not *ever* touch your students.

- Do not engage in extracurricular activities with students—dates, visits to bars, going to concerts, visits to museums. You might consider going to coffee with a group of students; it is probably a mistake to go with just one.

- Do not make suggestive remarks to students; do not give them meaningful or prolonged looks.

I really don't want to make it sound as though teaching these days is like navigating a minefield. For the most part it is a pleasure. But there are cautions that one must exercise in order to steer clear of dangerous waters.

3.9   First Kill All the Lawyers

When the bard first penned those words, he could hardly have anticipated the contentious and litigious society that has evolved. You never know who may sue whom for what. The first piece of advice that my financial planner ever gave me was to purchase a $1 million umbrella liability policy—just to protect against unexpected lawsuits. It costs about $150 per year, and seems to be money well-spent.

In the last section we offered some advice about sexual harassment issues. Another wrinkle of life that we must deal with these days is political correctness. Political correctness is a concept with a long history—dating even to a U.S. Supreme Court decision in 1793. Some people attribute the modern interest in political correctness to a 1992 paper of Arizona State University Law

---

[12]In fact many colleges and universities, especially public institutions, expressly forbid that you have such furniture in your office. Exceptions can be made for medical exigencies.

Professor Charles Richard Calleros [CAL]. The idea of political correctness is that language can be used to classify and oppress people. Even without the speaker being fully aware of the implications of his/her speech, he/she could be inadvertently fueling the repression of a class of people. Even calling your eighteen-year-old freshmen "girls" and "boys" is indicative of a bad attitude, and a lack of respect.

Of course colleges and universities are places that prize the exploration of new ideas; and often people latch on to these ideas and endeavor to turn them into a campaign or a way of life. This is what has happened with political correctness. Even some very prestigious, and otherwise quite scholarly and tony Eastern private universities have actually circulated to their faculties lists of terms that are considered to be "politically incorrect" along with suggested alternative terms that are more acceptable. Some might find it demeaning to be so instructed by their university administration. Others might find it laughable. One does not often hear of lawsuits that derive from a careless use of language, but there is a certain amount of malice and repression that derives from implementation of the idea of political correctness.

Of course we should all be respectful of other people—no matter what their ethnic background or social status. As a result, we should be careful of what we say. If someone cites you for being politically incorrect, thank them for the information and promise to do better. There is hardly anything more that a reasonable person can do.

3.10 The Two-Body Problem

Life in the 1950s was simple. Most math departments wouldn't hire women—period. Misogyny and discrimination were, quite frankly, rampant. Jobs were few. And nobody could be bothered to worry about social issues. Also salaries were low and the workload (i.e., the teaching load) was high. You really had to love mathematics to want to go into this profession. And your spouse had to love *you*.

It was like being in the Army: You went where the work was, you were paid a pittance, you lived in very humble circumstances. Your wife was expected to stay home, cook dinners (for you and for visitors), put up guests, go to boring colloquium parties and laugh at the Chair's lame jokes, and drink coffee with the other wives. Oh, yes, and raise the kids.

Things are different now. Your colleagues are both men and women.

Spouses are as likely to be husbands as wives. Most spouses have careers of their own. And, since college and graduate school are likely places to meet a spouse, many spouses are also mathematicians—or at least academics. This means that a mathematician who is looking to change jobs brings with him/her a certain amount of baggage—like a partner who wants a job too. This is called the *two-body problem.*[13]

There are certain legalities that prevent the recruiting Chair from asking many questions about this so-called "two-body" problem. But the candidate can bring it up, and probably he/she should. Because certainly this problem is bound to affect whether the candidate can accept any offer.

As mentioned elsewhere in this tract, there are some universities that have the power—by way of explicit edict of the institution—to simply cook up a job for the partner. And I don't mean just any old job; I mean a tenure-track faculty position. Other schools, like my own, will have a person in the administration who is the go-to person for spouses. This individual will provide orientation information for the spouse, professional contacts to aid in looking for a position, and can possibly even help to set up interviews. In many cases the burden will fall largely on the department Chair to try to scare up something for the spouse. This is not so easy, and will depend a good deal on whom the Chair knows, what kind of networking he/she has done, what sorts of strings he/she can pull.[14]

Some math departments, with the explicit blessing and support of their administrations, can try to make an offer both to the original candidate and to the spouse (assuming that the spouse is also a mathematician). Unfortunately it can happen that the original candidate is the person whom they really want—the one with all the credentials, the hot new theorems, the glow-in-the-dark letters of recommendation—and the spouse is somewhat less stellar. In that case the department must examine its collective conscience carefully and decide whether it wants to compromise its standards in order to gain a whole that is greater than the sum of its parts.

---

[13]The name derives from a classical problem in mechanics. Johannes Kepler and Isaac Newton completely described the motion of two planets acting on each other with the force of gravity. So the two-body problem is in effect a done deal. The three-body problem is still open. Well, that is the situation in mechanics. The two- and three-body problems in human relations are still difficult issues.

[14]At the risk of belaboring the obvious, it may be worth noting that it is often easier for a *non-academic* spouse to find gainful employment—at least if the university is in or near a city.

What people in the math department would like to convince themselves of—and I say this because it comes up every time that questions like this are considered—is that the position for the spouse is a "freebee." That the Dean is just giving this slot to the department, and it doesn't count in the great reckoning of how many tenure-track faculty there ought to be in the math department. Don't kid yourself. Governor Jerry Brown of California was right when he said that there is no free lunch; there are also no free faculty positions. Even if the Dean *says* it's free, let me assure you that it's not. As soon as the next Dean comes in, he/she is going to count positions in the math department by taking off his/her shoes and enumerating in the usual fashion. No slot will be treated any differently from any other.

If you, as Chair, are asked to handle a two-body situation, then my advice is to proceed with some delicacy. Show the candidates the utmost consideration, and be solicitous of their needs. Do what you can do for them.[15] You may be thinking, in the back of your cranium, that this is not how you were treated when you were a candidate here (some thirty or more years ago). *Your* spouse was given no particular consideration. Why is this candidate so special? The answer is that this candidate is *not* so special. But the rules of engagement have changed. If you want to compete in this marketplace, then you have to come to terms with these new exigencies.

---

[15]There are many layers of complexity here. If both spousal candidates are in the *same department*—namely yours—then you have more control over the situation and can perhaps work something out (assuming, that is, that your department and your Dean want to go along with it). If the spouses are split across two departments, then you are dealing with two value systems and two different *modus operandi* and two different sets of scholars. Things are even more difficult if one of the two spousal candidates is clearly the star and one is not. You may be able to sell a star and a sub-star together to your own department. It is unlikely that you could sell a star to your own department and a sub-star to some other department.

Of course there is also the situation where another department is trying to hire some superstar and wants *you*, the math department to hire the spouse. Likely as not, if this is not a candidate of your explicit choosing, it is probably not somebody that you want all that badly. Will the Dean provide incentive—either pecuniary or in some other tangible form—that will make it attractive for you to play along?

Part II

Living the Life

Chapter 4

Research

The profession of film director can and should be such a high and precious one, that no man aspiring to it can disregard any knowledge that will make him a better film director or human being.

Sergei Eisenstein (filmmaker)

He has made a profession out of a business and an art out of a profession.

Clifton Paul Fadiman (literary critic)

And to think that merely because I was a professional baseball player, I could ignore what was going on outside the walls of Busch Stadium was truly hypocrisy and now I found that all of those rights that these great Americans were dying for, I didn't have in my own profession.

Curt Flood (baseball player)

To be a poet is a condition, not a profession.

Robert Frost (poet)

Journalism is the only profession explicitly protected by the U.S. Constitution, because journalists are supposed to be the check and balance on government. We're supposed to be holding those in power accountable. We're not supposed to be their megaphone. That's what the corporate media have become.

Amy Goodman (journalist)

I am sufficiently convinced already that the members of a profession know their own calling better than anyone else can know it.

Asa Gray (botanist)

Only two things are infinite, the universe and human stupidity, and I'm not sure about the former.

Albert Einstein (physicist)

In the end, we will remember not the words of our enemies, but the silence of our friends.

Martin Luther King (political leader and minister)

4.1   What Is Mathematical Research?

The trouble with education is that it can make you too dependent on your teachers. Breaking away from your Ph.D. advisor and your other mentors and charting your own course in the academic or professional world is actually quite difficult to do. In point of fact most people do *not* succeed in establishing their own scholarly identity.

One of the—although certainly not the only—means by which we establish our independent *Gestalt* and *Weltanschauung* as mathematicians is by founding and developing an individual research program. If you are fresh out of graduate school, then it is probably not at all clear to you how this is done. Likely as not, your thesis advisor gave you a thesis problem and helped you to solve it (although the road can be a little more complicated than that—see [KRA7]). This is not at all what real life is like. You are now an Assistant Professor in a math department, you no longer have a thesis advisor, you do not have anyone suggesting problems to you, and you certainly don't have anyone holding your hand while you are engaged in your research. So where do you turn? What do you do?

This is a situation in which omphaloskepsis will not do the trick (although it has been frequently tried). You cannot succeed at anything in life unless you know precisely what it is that you are trying to accomplish. So let us enunciate that first: You are trying to establish yourself as a player in an area of mathematical research. You do so by proving some good theorems and publishing some papers that report on those theorems.

There are several potential roadblocks to your progress on this program:

- You need to find problems that are of interest to you, will give you the fire in the guts, will so consume you that you must solve them.

- The problems of the first stripe must also be problems that fit naturally into the current ongoing flow of mathematics as it is practiced today. There are many fine mathematicians throughout the world who maintain important and distinguished research programs. They, in effect, *define* what mathematics today is all about. Your research program should mesh with what is of interest to the leaders in the field. Because, in the end, it is they who will judge the value of your work.[1] If

---

[1]They do so in a variety of ways: By refereeing your papers, by refereeing grant proposals, by writing letters for tenure and promotion cases, and by organizing conferences and issuing invitations thereto.

you decide to write a sequence of papers about some obscure topic that is out of the mainstream of modern mathematical thought, then you may derive a certain amount of personal satisfaction from the process but you will not establish a reputation and you will probably not get tenure.

- You need to find problems that you have a chance of solving, or at least of making a significant contribution toward their understanding. This is a tricky point, because it takes considerable experience to know how to weigh a problem and to determine whether this is a problem that you can make a dent in. That is why you must go to seminars, go to conferences, and *talk* to people. One of the great things about modern mathematics is that there is so much free and open communication. Very few of us keep our problems secret, or keep our research programs under wraps. Even going to people's Web pages can be an immensely informative and rewarding experience. Many people maintain problem lists, `Wiki`-pages, chat rooms, and other OnLine devices for helping people get up to speed in the field. You should take advantage of all these tools.

- You need to actually get the papers written—and written in a form and style that is of publication quality (for guidance here, see [KRA2] and [STE]).

- You need to get the papers submitted to journals, get them through the editorial process, and get them accepted.

These five—especially the first three—are perhaps the most fundamental issues in getting a research program going. Unfortunately there are a number of other junctures where you could stumble or fall. There are quite a few very talented mathematicians in our world who just cannot seem to manage to get their work written up. Or, having written it down in *some form*, they cannot seem to find the energy or discipline to put it into a form that can be submitted to a journal. Or having perhaps found the wherewithal to get the paper sent to a journal, they do not have the maturity and the tenacity to deal with the editor and the referees and make the edits and corrections that have been requested.

Let me stress that *this is your life*. If you choose to be an academic mathematician, then you will spend a good deal of your time—for the next forty years or more—wrestling with getting your ideas recorded, wrestling with getting your ideas into publishable form, grappling with editors, and going three falls with referees. It is *very* hard work, and requires considerable discipline and commitment and often some humility. Not all of us are cut out for it. But there you have it; this is the scholarly world.

4.2   How to Do Mathematical Research

My teacher and friend Fred Almgren used to say that a graduate student should spend four hours every day doing mathematics; what he/she did with the rest of the day was his/her business. Now one must understand that Fred had incredible powers of concentration. When he did mathematics, he was not distracted by music or coffee cups or phone calls or e-mail or answering the door or anything else. All he did was sit and think. Absolutely nothing could distract him. Fred was often twenty minutes late for class because he was so lost in thought that he had no sense of time.

Not many of us are blessed with such powers. But it is an observed fact that it is nearly impossible to do serious mathematical research for more than four or five hours in a day. You can kid yourself and pretend to spend twenty hours a day working—hanging out in the library, sitting in the coffee room BS-ing about mathematics, going OnLine and reading chat rooms and Websites of famous mathematicians, staring at printouts of the latest papers from famous guys at Princeton and Harvard. But the fact is that, if you have the discipline and stamina to do so, you are much better off doing what Fred Almgren did: Go some place where you absolutely, positively cannot be interrupted and sit in front of a tablet of paper and think hard and calculate.[2] Try things. Scribble and conjecture and fiddle and tinker and make mistakes and try to learn from them. I've had great ideas while driving the car or taking a shower or lying on the lawn staring at the birds. But my most productive times have been times spent alone in front of the old tablet with a pen clutched in my fist, calculating away. If you can produce four or five productive hours like this five or six days per week, then you will be headed in the direction of being a successful mathematician.

Keep a notebook. Gauss did. Riemann did. In modern times, Lars

---

[2]Lars Ahlfors liked to say that all a mathematician needs in order to do his/her work are paper, a pencil, and a wastebasket. He went further to say that a philosopher required even less equipment, as he/she had no need for the wastebasket.

Hörmander would sit down at the end of each day and record (in his note-book!) what he had tried and what had worked and what had failed and what he had learned from the process. Even if you never look back at any-thing you have recorded in this notebook, it is good discipline to assess each day as it passes.

Now fair is fair and we must acknowledge that the shape and nature of mathematical research is ever-changing. Some people—not just numerical analysts but modelers and many others—do much of their research on the computer. Some people do their research in teams. *Interacting with other people* is a key part of their research program. Such people will spend a few hours every day just banging their heads together, generating ideas. Some people actually build physical models, and derive ideas from them. David Hoffman and his team GANG (Geometry, Analysis, Numerics, and Graphics) at the University of Massachusetts would generate numerical solutions of the minimal surface equation and thereby create graphical images on the computer screen of certain minimal surfaces that nobody had known even existed. They would stare at these graphical images, discuss and debate them, and draw certain conceptual conclusions from what they saw. *Then they would sit down and prove theorems in the traditional fashion with pen and paper.* I myself do research in collaboration with some plastic surgeons. At least some of our work is fueled by activity in the operating theater (performed by physicians), opening up patients and wielding the old scalpel.

It should be clearly understood that applied mathematicians—just to take an example—do not necessarily do mathematics in the traditional fash-ion. They do not necessarily prove theorems. A big part of what they do is *modeling*; finding a mathematical structure that describes the world around us is a major achievement—for them and for science. They also do numer-ical calculations and graphical simulations. Applied mathematician Stanley Osher once said in an invited lecture that, "Those other guys didn't get any-where with this problem because they were trying to prove a theorem. We just wanted results." One upshot of these considerations is that the applied mathematician's approach to his/her work could be a bit different from what I have described above.

Some people, as part of their research program, read everything that they can get their hands on. Some people read almost nothing—preferring to generate the ideas themselves from whole cloth. This is clearly a personal matter. After a while reading becomes cloying—it is certainly *not* like *doing*. But it *is* a way to immerse yourself in the subject, and can be a way of generating ideas. You must decide for yourself what gets your juices flowing.

Like teaching, doing mathematical research is an intensely personal activity. Part of your job, when you are starting out in this discipline, is to figure out what works for you, to develop the necessary tools and talents, and to put them to work. Along the way, you will falter and fail and have to pick yourself up and try again. But this is life. Progress will take the form of two steps forward, one step back. But there will be progress indeed if only you are committed to the process and have your goals clearly in sight.

The flip side of what I have been saying here is that you must have nontrivial, contiguous chunks of time in which to do your work. Doing mathematics is not like polishing shoes—you cannot do your work in twenty-minute snatches. In fact it frequently takes an hour to get all the ideas organized in your head *before* you can actually start to calculate and think and try things. And you must have the leisure to turn ideas over in your mind, try different approaches, try to prove something is true by first trying to prove that it is false. The great and mysterious fact about mathematical research is that you frequently *do not know what it is you are looking for*. That makes the quest exciting, but it also makes it frustrating and time-consuming. On the one hand, you cannot productively spend twenty hours per day doing serious mathematics. But you also cannot usefully spend just twenty minutes per day doing serious mathematics.

It will also happen that you prove a good theorem, give some seminar talks on it, garner praise and adulation, write up the result so that it glows in the dark, submit it to a journal, and have it rejected. Be of stout heart. Henrik Ibsen and Jane Austen and many another fine writer[3] had his/her best work rejected. Some of my own finest and most influential papers were not only rejected at first, but were wholeheartedly insulted by the referees. And the editors were of no help either. In the intellectual world, tenacity and self-confidence are two of the most important attributes that you can have. Maintain your modesty and your dignity, but cultivate these two essential properties in yourself.

If a paper is rejected, read the reports carefully and endeavor to learn from them. Find out how to make the paper better and do so. Show the paper to a senior mentor or even to a peer colleague. Do everything you can

---

[3]In fact Alfred A. Knopf—arguably the gold standard among American publishers— recently made the *New York Times* for having turned down Vladimir Nabokov (*Lolita*), Jack Kerouac (*On the Road*), *The Diary of Anne Frank*, Pearl S. Buck (*The Good Earth*), George Orwell (*Animal Farm*), Isaac Bashevis Singer, Anaïs Nin, and Sylvia Plath.

to capitalize on the situation and to learn something from it. Then resubmit and see whether you fare better.[4] I have never had a paper that I could not get published. But I have stumbled, I have had papers rejected, I have needed to swallow my pride and work hard to shore up a paper before I could get it out the door. Sometimes I've had to publish in a journal that was a slice below my original goals. That is part of the publishing game; you will get used to it. And you will learn from it. In the end it is rewarding, and it will make you stronger.

4.3   How to Establish a Research Reputation

This has to be the key question, as it is the nub of success in the academic world. It is not just enough to write things and publish them. You must get people to *read* them and appreciate them. In the best of all possible worlds, you want people to be running seminars in which they discuss your work. If you write a book, you want people to read your book and talk about it. You want fellow scholars to feel that your ideas have an impact.

This is a lot like asking how to become a great violinist or a great painter. It is not enough to learn the hand positions and master the details of a Bach partita (for music). Nor is it enough to learn the basics of color and composition (for art). There is a great component of inspiration and, indeed, of genius. Either you've got it or you don't.

But please do not conclude that I am saying to you, "If you have to ask, then you haven't got it. Better ply your wares elsewhere." There is more to it than that. Certainly being smart is a big part of being a mathematician. And being inspired is an inestimable commodity. But the single most important feature of a successful mathematician is a capacity for hard work.[5] And dedication and tenacity. Plus the inner strength to forge on even when the going gets tough and things do not look very promising. Note once again

---

[4]It is the rare journal indeed that will want you to resubmit a paper that has been rejected *again to that same journal.* Many journals say explicitly in the letter of rejection, "We wish you good luck with your paper elsewhere." Even if the editor does not make this outright statement, it is best to assume that you will do better to submit your paper to another journal.

[5]Even the most brilliant and inspired mathematicians that I know—Fields Medalists and members of the National Academy—have a very strict work ethic and a dedicated routine for doing mathematics. They put in a good many hours on mathematics each and every day.

that all these attributes are important because they support your ability to work hard and to dedicate yourself to mathematics.

Most mathematicians, indeed most academics, are quite fragile. They come from a lifetime of many successes and very few failures. School is, quite frankly, rather rote and repetitious. Once you get the hang of it you can do quite well—even through graduate school. But once you are a professional academic you are in the big leagues. You are competing against people who are every bit as talented as you and sometimes more so. Now you will not triumph every time. Far from it. Many academics fall by the wayside because they are simply not temperamentally fit to deal with failure or disappointment. Even small disappointments. If you can muster the tenacity to pick yourself up and fight on, then you will have a great advantage over most of your colleagues.

You can write very good papers—perhaps not the papers that will reinvent a subject, or start a whole new direction—that make a solid contribution to a field just by working hard, mastering a body of material, doing lots of calculations, thinking deeply, and reporting what you have learned. This can be the basis for a respectable and rewarding career. And this is in fact what most of us do. You may not win the Fields Medal, or be a Plenary Speaker at the International Congress, but you can hold your head up and feel that you have contributed to the infrastructure of your field. You have made a contribution that is appreciated and will be part of the canon.

We are not by nature people who advertise ourselves.[6] But there are sensible things that you can do to establish your reputation. Giving lots of talks about your stuff is certainly a good way to get some attention for your ideas. Writing expository articles is another (although you do not want to develop the reputation of being someone who writes *primarily* expository articles). And just talking to people—sharing your ideas—is another useful way to cast your bread upon the waters. Also go to conferences. Make sure that your efforts are known to the experts in the field. Invite people to your department and get to know them. Tell them about your research. In the end, people will form their own judgments. It is the discipline itself that decides the value of your work.

---

[6]When Carl Sagan got his first job at the Jet Propulsion Laboratory in Pasadena, one of his first acts was to go out and hire a publicist. This is not typical behavior for an academic.

4.4 Seminars

Seminars are a key part of the mathematical life. A seminar is a device for sharing new mathematics with colleagues, for stimulating useful discussions, and even for spawning new collaborations.

If you are a skeptic, you may say that if you want to share mathematics with a colleague, then you will simply knock on his/her door and start talking about mathematics. But it is a fact of life that people often need a catalyst to get their communication started. The seminar is a useful device for that purpose. It is also specifically designed for getting graduate students involved and for getting young faculty jump-started on their research careers.

Be sure to find out about the relevant seminars in your department (or in departments at nearby institutions).[7] Be an active participant. Give talks in the seminar. These talks need not be about the latest theorem that you proved last night. They could be about an interesting paper that you read recently, or an interesting idea that looks potentially fruitful but hasn't paid off yet. A seminar *should* be a friendly, generous, constructive exchange of information.[8] It is a positive way to socialize (in a professional manner) with your colleagues. It is a way to get to know people.

You will no doubt reach a stage in your career when it is appropriate for you to start and to run a seminar (or perhaps to assume the helm of an already existing seminar).[9] It doesn't take much to do so. You hang

---

[7]Here "nearby" could mean within a one-hundred-mile radius. A great many mathematicians have taken jobs around the Boston area just so that they can attend seminars at Harvard, Brandeis, and MIT (to name a few). When I was at Penn State, people used to drive a considerable distance from branch campuses to participate in our seminars.

[8]Penrose's "Friday meetings" at Oxford had a rule that any graduate student could ask any question, and it *had* to be answered. No disparaging comments or put-downs were allowed. Phil Griffiths's "Nothing Seminar" at Princeton was similar.

[9]In France, the seminar of Cartan and the seminar of Lelong—to name just two—exerted an enormous influence over the development of twentieth-century French mathematics. The Gelfand seminars in Moscow played a pivotal role in shaping modern Soviet mathematics, and mathematics in general. We do not have such a system in this country. But there are some notable individual examples: Walter Rudin's seminar at the University of Wisconsin is widely viewed as having been extremely influential, and important in the development of many fine analysts. Phil Curtis's seminar at UCLA and Paul Halmos's SCFAS (Southern California Functional Analysis Seminar) also had a powerful effect. Irving Kaplansky at Chicago (algebra) and Norman Steenrod at Princeton (topology) and S. S. Chern at Berkeley (differential geometry) had influential seminars. David Vogan's representation theory seminar at MIT is highly influential. There are surely examples

an announcement on the departmental Bulletin Board (and these days you might also send out an *e*-mail and put up a Web page) enunciating what you want this seminar to be about and why there are compelling reasons to start something up. Then you arrange for an organizational meeting.

If you've done things right, and chosen a hot and lively topic (it's a good idea to knock on some doors and talk things up with some key potential participants), then a bunch of people will show up at your organizational meeting. At that point you should engender a lively discussion of what this seminar should be, or what people want it to be. And then you need to sign up some speakers. If people have shown up at this meeting, then they will know that they are liable to be tapped to be speakers. So you certainly ought to get at least a handful of volunteers. You can definitely offer to speak yourself, and you should.[10] And then you are off and running. You must consult with the appropriate staff member to line up a suitable weekly time slot and of course a room. And then you have a seminar.

Even though we don't have the French system, it is really true that a good seminar can have a lasting effect, and can play a decisive role in the development of mathematicians both young and old. The seminar that Dick Palais ran in the 1960s on the Atiyah-Singer Index Theorem has had a tremendous effect. It certainly helped the world to understand what this important theorem is all about. It helped to promulgate and establish the significance of pseudodifferential operators and K-theory. And the volume that Palais published based on the seminar established his presence in the mathematical firmament.

And this last is a notable point. Your seminar *could* in principle publish periodic volumes of its proceedings. At least you could publish them on the Web. And if you can develop a proactive relationship with a traditional publisher, then perhaps you can publish them in the more traditional fashion. As an instance, the *Cabal Seminar* publications out of UCLA and Cal Tech have had a positive and lasting effect on the logic community.

Running a seminar makes you a player in the mathematical life of your department. When a job candidate or a visitor comes through, and asks, "What goes on here?" then it is natural for the host or the Chair to say, "We have this seminar run by A and we have that seminar run by B." These

---

from all parts of mathematics.

[10]Often the organizer is anticipated to be the first speaker. After all, he/she should set the tone and direction for the seminar.

are bellwethers of what is happening in the department. And the Dean likes to know that there are seminars. These are indications that the department has a lively scholarly life.

4.5 Writing Papers

A detailed account of how to write a paper may be found in [KRA2]. See also [KRA5]. Here I shall only hit some of the high points.

There are several reasons to write a mathematical paper:

1. You want to stake your claim to a subject area. You want to establish, for the historical record, that you are the one who discovered this theorem and you proved it first.

2. You want to teach something important to the mathematical community.

3. You want to establish and describe an important new body of techniques in the subject.

4. You want to explain something "known" that has never been adequately explained or understood before.

5. You want to exposit an important area of mathematics, teach it to a broad audience, change it from a disconnected collection of isolated results into a coherent field.

If your only goal is **1.**—and I am sorry to say that for many mathematicians that *is* the only goal—then you probably don't care much about the quality of writing in your paper, you probably don't care who your audience is, you probably don't give a lot of thought to whether you are communicating effectively or well, and you certainly don't need to read any further what I have to say in the matter. If instead your goal is to communicate, to enlighten, and to play a dynamic role in your discipline, then read on.

The first and most important precept of any sort of writing is that you must have something to say. In mathematics—much more so than in many other academic disciplines—this is really a cut-and-dried matter. Either you have proved a new theorem or you have not. It is virtually impossible in mathematics to write a paper (and to get it published!) that says, "This

morning I woke up and tried to prove the Riemann hypothesis and I failed."
Dress it up as you will, this just isn't going to fly. The necessary and sufficient
condition for being able to write a new paper (assuming that this is not an
expository paper) is that you have a non-trivial new result to present.

So let us suppose that you do have something new and exciting. The
next prerequisite for effective writing is that you consider carefully who your
audience is. Is it graduate students? Undergraduates? Fully vested mathematicians? Researchers? Specialists in your subject area? Super-specialists
on this particular problem?

In most cases, if you have proved a new theorem in your particular area
of research, then that theorem is probably fairly specialized. Probably quite
technical. Likely of interest only to a limited audience. So your target for the
paper you are writing is going to be *at most* specialists in analytic number
theory, or specialists in several complex variables, or specialists in nonlinear
partial differential equations of second order. It could be more special than
that. Many a good paper addresses a limited group of people who have
actually spent time on that particular problem.

The considerations in the last paragraph will affect how your paper is
written. How much can you assume that your audience knows? Will that
audience be acquainted with all the technical jargon? Will it be familiar
with all the history and background of the problem? Will it have all the
motivation in place? Will it just be saying, "OK, what's the statement of
the theorem? Where's the proof? Save the hearts and flowers for another
occasion." Clearly you cannot formulate your paper effectively (in your mind,
before you set pen to paper) until you have these factors precisely in focus.

Your new paper should be well-organized. It should begin with an Abstract. Most journals require one, but it's a good habit to always formulate a
brief, pithy, eye-catching Abstract. Then proceed with some background and
motivation for the problem at hand. It can safely be said that *most readers*
will read your abstract, and *many readers* will read your introduction. The
number that will forge on beyond that is small. So those first two components of the paper are of utmost importance. You should polish them to a
high sheen, and fashion them so that any well-informed mathematician can
get something of value from them. Tell people what is in the paper and why
it is interesting. *Give them a reason* to want to read on.

The introduction need not be prolix. One page is more than enough—less
if you can point the reader to a good and accessible source that *will* give all
the needed background. It is nice if you can give an informal, less technical,

rendition of the key result before you get down and start rolling in the mud. That way the reader can spend about ten minutes and get a good idea of **(i)** what this paper is about, **(ii)** what background is needed to appreciate the paper, **(iii)** what is the main result, **(iv)** whether or not he/she wants to put in some time and try to understand all the gory details.

The next part of the paper could give some notation and statements of key supporting ideas. Although this material is in principle somewhat technical, mathematicians find it rather comforting. This is our bread and butter, and this is how you will bring the reader up to speed. You may follow this section with a section that (building on the material just described) gives careful and rigorous statements of the main results.

If the proofs that you are going to present are difficult and tricky, then it is nice to provide a section that outlines the key steps of the proof. This would be silly if your full-blown proof is just 2.5 pages and not too hard to read. But if the proof is long and complicated, then an overview is much appreciated and will help with the difficult reading that lies ahead.

Finally you present all the details of the proof. Do take advantage of the formalism of modern mathematics to help you lay things out. Have enunciated lemmas, propositions, sublemmas, and so forth to organize the ideas and break the reasoning up into steps. Provide some intermediary prose to remind the reader of where he/she has been and where you are going. If appropriate, let the proof span several sections so that the layout of the ideas—in the large—becomes clearer. Always be sure that the reader knows where he/she is and what is being proved at any given moment. Do *not* have a lot of hanging lemmas and unproved claims and unformulated hypotheses. It is always best to have everything up-front and explicit.

It is also good exposition to avoid making universal statements like, "From now on all spaces are paracompact and semi-metric." or "The letter X will henceforth be a sub-Banach space with a tensored lattice structure." It is difficult for the reader to remember such declarations, and certainly not much trouble to repeat them as needed. Anything that causes the reader to stumble or hesitate is not good. Repetition is the most effective teacher that there is.

It is OK, and commonly done, to end a paper with something like

> . . . and therefore the sequence converges hemi-semi-demi uniformly
> at almost every point of the pseudo-set. □

This communicates a certain *joie de vivre*, and lets the reader know that you are an author who knows when to call it quits. But such an ending is also quite likely to leave the reader saying, *"Huh?"* Because he/she will have to do a double take, review some of the definitions, go back and figure out why this convergence finishes the proof and establishes what was claimed, and satisfy him/herself that the job has been done.

It is always nicer, and certainly more effective communication, to make some concluding remarks that draw all your ideas together, point out what has been accomplished and how, and suggest further avenues of research. These remarks can be contained in a special section called *Concluding Remarks*. Just as a good talk should end with some summary observations, so should a good paper.

4.6   Writing Books

I have written a number of research papers, some of them rather good. And some of them quite influential. But my work that has had the greatest and broadest and most lasting impact is the books that I have written. A book is a different type of writing, with a different sort of audience, and certainly different goals.

I speak at the moment not of textbooks (which will be discussed elsewhere) but of stand-alone research monographs and the like. A monograph can be your definitive statement about your subject area, based on your struggles to master it and to make definitive contributions to it. You can provide the genesis and history of the ideas, the motivation for the key concepts, a description of the important techniques, and a panorama of the applications.

This is definitely the way to make your mark as a professional mathematician. Especially if the topic has not been exposited before, your book can make a notable change in the infrastructure of mathematics. A good book provides a synthesis of ideas, and presents them from a particular point of view. In writing this tract you shape your subject in a powerful and fruitful manner. A treatise such as I am discussing here will receive much wider attention than a research article and be accessible to a vast cross section of students and professional mathematicians. Your book will stand for a long time, will be read by a good many students and professional mathematicians, and will help people to understand a part of the world.

A good deal has been said about the particular tools and techniques that should be used to write a good book (see [KRA2]), and I shall not regurgitate them all here. Let me just say for now that book writing is not for everyone. Whereas writing up a research paper could take a month or two, the writing of a book could take one or several years. A book is perhaps fifteen times as long as a paper, but it could take fifty times as much of your life. It is a complicated business. There are terrific organizational issues, and delicate decisions about what to include and what not to include. In addition, the process is sufficiently protracted that there could be fast-breaking new results that change the nature of the beast even while you are writing the book![11]

Antoni Zygmund's famous book [ZYG] is a monument to Fourier analysis and classical analysis in general. It may be the most important analysis book of the twentieth century. But Zygmund himself said that he gave up writing thirty papers in order to pen the volume. You may want to do a similar estimation *before* you sit down to write your own book. How will this project impact your career? Will the overall effect be favorable? If you are a significant person in your field and you let your colleagues around the world know that you are setting time aside to write *the* book in your subject, then it is likely that they will be very supportive. They may even help. But you are the person who has to carry the weight. It is a long haul and you must stay the course to the bitter end. You don't benefit anyone if you don't finish the project and end up with a stack of notes sitting on your shelf. Talk this project up with your trusted friends and advisors before you ever set pen to paper (or fingers to keyboard, as the case may be).

On top of all that, you must gird your loins and figure out how to deal with a publisher. This is really not at all like dealing with a journal—which is primarily doing business with fellow academics. Often when you publish a book you are dealing with a commercial publisher. Publishing is a business, and business people have methods and values which may not be familiar or comfortable. You must talk about scheduling, copyediting, royalties, and a

---

[11]These days it is quite common to use the Web to keep your book up to date. That way you need not worry that important new theorems are being proved even while your book is in press. And you can also post ancillary graphics, extra examples, nonstandard exercises, animated diagrams, and a variety of other resources. Often your publisher will help you to develop a Website for your book. In many instances the publisher will insist on it.

variety of other issues in a manner which some people find confusing and others find offensive. The book [KRA4] can give you some guidance in the matter.

Writing a book is rewarding and can be important. But it can take a lot out of your life, and in fact can even alter your life. It can affect your personal relationships, particularly the amount of time that you have available to spend with your family. You should think carefully before you embark on this road. Time management is also a crucial part of the process. If you cannot manage your time in the small, you will have a much harder time getting your act together with a project as big as a book. Far too often does it happen that a book gets started with a great flush of energy and enthusiasm, but ultimately it peters out and never gets completed (or published).[12] That is really too bad, and also a waste of time and energy. You are much better off to know what you are getting into and to plan how you will execute and complete the project—before you ever set pen to paper.

You also should be aware that book writing is not necessarily rewarded in the same way that paper writing is. Generally speaking, book writing is not seen as doing research. True, André Weil's *Basic Number Theory* [WEI] treats the Riemann hypothesis for function fields over finite fields. The recent book by Harris and Taylor [HAT] proves the local Langlands conjecture. But these are the exceptions rather than the rule. Books tend to be exposition, and exposition is valuable. But it is not proving theorems. And your Chair and your Dean know it.[13]

4.7 Working on Your Own

The default mode of work in the academic mathematics profession is to labor away in isolation. You are the single-combat warrior. It is you against some

---

[12]It is *much harder* to finish a book than to start one. When you begin the task, you are full of energy and ideas. You will have a ball setting your view of the world to paper, shaping the subject in your own fashion, telling the story as only you see it. Much harder is engaging in the tedium of polishing the book to a finished product, producing the Index and the Bibliography, checking the accuracy of all the details.

[13]I would be remiss not to observe that different institutions have different value systems. The comprehensive universities view scholarship in perhaps a different manner than does Princeton or Harvard. If a comprehensive university faculty member writes a textbook, then that is considered to be a notable achievement, and it is duly rewarded. Publication of papers is too, of course.

large problem, or group of problems. If you win—if you defeat the problem—then the victory is all yours and you can revel (however briefly) in the glory. You are like David defeating Goliath or Siegfried defeating Fafnir the dragon. Likewise, if you lose, then you are nothing. You have nothing achieved, and have nothing to show.

It is a tough life to cut out for yourself, with potential big wins and possibly catastrophic losses. You can not only fail to solve your problem, you could also lose your peace of mind, your mental stability, your friends, and your marriage. I kid you not: it is a zero-one game in the strongest sense of the word. Look what happened to John Nash [NAS]. Or Grigori Perelman [NAG].

But there is another way to look at the matter. All the best mathematicians that I know are masters at snatching victory from the jaws of defeat. From a distance, it may *look* as though they bound from success to success, solving one big problem after another. This really is not true. Like most of us they end up sticking most of their work in the filing cabinet or the trash bin. But they have a specially developed skill for looking at a heap of calculations, and six months of trials and errors, and eliciting something of value from it. A weak or uncertain mind would scoop up all the paper and dump it in the circular file. A more creative mind will think things over and say, "Well, I didn't solve the big problem, but look at what I've learned. There are some new connections here, and that elucidates this other point." And before you know it he/she has written a terrific paper.

The advantage of working on your own, the joy of it, is that you can do your work whenever you like—sitting in your office, or going for a walk, or playing pool, or swimming, or mowing the lawn. You can work at your own pace, and explore any diversions or byways that strike your fancy. Who knows what gold you may strike?

The disadvantage is that you could easily become discouraged, demoralized, depressed. You could decide to give up, change careers, take a different job. These days almost everyone collaborates, and there is plenty of good reason to do so. Mathematics has become large and unwieldy. The problems are immense, and can be overwhelming. It is a great help to have a sounding board, a reality check, a verifier that only another person can be.

Even if you don't collaborate, you should treat mathematics as a participatory endeavor. Talk to people. Go to seminars. Participate in conferences. Attend the International Congress of Mathematicians. Publish papers. Read the *Notices*. Play the game. It is comforting to be part of the process, and

part of the life. It will keep you strong and focused, and help you to move
ahead.

4.8   Working in Collaboration

One of the joys of modern scholarly life in mathematics is writing papers with
your friends and colleagues. Eighty years ago, when Hardy and Littlewood
were producing their myriad joint papers, such collaboration was virtually
unheard of. Now it is the norm.

Much of modern mathematics is cross-disciplinary—a marriage of analysis
and geometry, or PDE and Lie groups, or logic and algebra (to name just
a few). This makes the subject richer and more diverse, but it also makes
it more complex. It is rather more difficult for a lone individual to make
any progress. Put another way, it is very natural for people from different
disciplines to get together to work on a problem that uses ideas from many
parts of mathematics.

I have written more than 150 papers, and at least half of them are joint.
Working with other mathematicians is an essential part of my professional
life. I generate new ideas naturally and with pleasure while drinking a beer
with a colleague, while having a "math rap session" at the blackboard, while
going for a hike, or in the debriefing after an interesting talk. Working alone,
it is easy to become discouraged and confused. Having a collaborator can
give you strength, give you someone off of whom you can bounce your ideas,
give you a regular re-centering of your course.

If you want your collaborations to proceed smoothly and salubriously, if
you want everyone to be happy at the end so that you can all look forward to
fortuitous future collaborations, then there are certain precepts of courtesy
and professionalism that ought to be followed:

- Usually the question of who will do what, and who will contribute what,
  is answered automatically as the ideas develop. Everyone should *really
  want to* be part of this team. Everyone should really want to pitch in.
  When a new task comes up, you don't expect to hear, "Oh, I've got to
  go do my gardening this week." Instead you expect to hear, "That's
  important. I'll write a draft and show it to the rest of you." or "That's
  the sort of calculation that I like to do. Let me try it."

- Nobody should worry about who gets credit for what. Mathematical collaborations are democratic endeavors, and all the participants are equals. The custom (unlike in other sciences) is to list the authors alphabetically. So there should be no squabbling over priorities or ranking. It often will be true that, on a given paper, the second author (named Zymurgy) made a greater contribution than the first author (named Aardvaark). But that is liable to be reversed in the next paper. In any event, it is counterproductive and damaging to the psyche to spend time worrying about these peccadillos. The point of collaboration is to contribute what you can and derive from it what you can. If at the end it turns out to be not entirely satisfactory, then everyone can shake hands and go his/her separate way. There are many other fish in the sea.

- Part of your job as a collaborator is to provide moral support (and praise, when appropriate) for your co-workers. Some of the work will be difficult and confusing, other parts will be long and tedious. You can take some pleasure, and learn something yourself in the process, by explaining some of the ideas to your teammates. Say a kind word when someone makes a nice observation, or contributes something of value. Lend a hand when and where appropriate.

- Everyone brings something different to the altar. Some are good at concepts, others at calculation. Some shine at writing the stuff up, others at posing the critical questions. It is essential that each of the collaborators respects and appreciates the other collaborators and what they have to give to the project.

- Everyone should bend over backwards to be courteous to everyone else. Everyone should avoid being critical or demanding or captious or divisive. This is a joint endeavor with a worthy purpose, and everyone's eyes should be focused on the goal.

You want to exercise some care in choosing your collaborators. I usually pick collaborators whom I like, with whom I have a lot in common, and who share my interests and goals. This takes place fairly naturally, and has led to many happy team efforts. I hope that you have a similar experience.

4.9   Publishing Papers

The books [KRA2] and [KRA4] also contain detailed discussions of how to get a paper published. This is a necessary part of the process, and not a trivial one. But it need not be an ordeal. I provide here a brief discussion (see also [KRA5]).

The advent of the Internet has changed everyone's view of what professional publishing is, and also of what it means. It has become quite simple to post your work on the Internet. For mathematicians these days the canonical place to put your paper is the preprint server `arXiv`. Originally created by Paul Ginsparg at Los Alamos and now hosted by Cornell University, this is where mathematicians from all fields put their work as soon as it is created. As of today, about 10% of all new mathematics papers are posted on `arXiv`, but the number is growing steadily. There are many other places to post papers, including departmental Websites and specialty Websites for particular subject areas like K-theory or linear algebraic groups. You can post your work on one preprint server or many. There are no restrictions.

The point of confusion that I wish to discuss here is whether putting your work on an Internet preprint server is *actually publishing it*. In some technical sense it certainly *is*. For you are setting your ideas before a broad audience (in fact, in principle, before the entire world) and opening yourself up for criticism and discussion. That is what the academic/scholarly life is all about. So it is good. Certainly when Nathan Seiberg and Ed Witten posted their ideas on the Web about what are now called the Seiberg/Witten equations, it caused a major upheaval in modern gauge theory. This is an example in which the Internet played a decisive role in making things happen quickly. For a couple of years theoretical physics was not the same.

But Seiberg and Witten ultimately published their ideas in a standard venue (the *Journal of High Energy Physics*)—that is to say, a recognized, refereed, scholarly journal. There are lots of good reasons for doing so. First of all, peer review is a form of validation for your work. When you pick up the latest issue of the *Annals of Mathematics* or *Acta Mathematica*, you know that the papers inside have been very carefully reviewed. Thus the time that you spend with these papers is likely to be worthwhile and rewarding. Perhaps most significantly, if you are up for tenure or for a promotion, or for some other signal form of recognition, then peer-reviewed journals are what counts. This is what the Dean sees as the ultimate filter for solid, respected academic work. If he/she is going to grant lifetime tenure to a candidate, then he/she is going to want to know that this candidate has published significant scholarly material that has been reviewed and validated and given a stamp of approval by other recognized scholars.

Put in other words, the good news about the Internet is that it gets your work out there rapidly, cheaply, and to a very broad audience. The bad news is that it sits in an undifferentiated sea of literally billions of electronic documents. We need some mechanism to determine what (mathematical and otherwise) is of value and what is perhaps safely passed over. You want your *Curriculum Vitae* to show quite plainly that you are a scholar of some accomplishment whose work is recognized and valued by the scholarly community. Traditional, peer-reviewed journals are the way to get this done.

There are also questions of archiving. Hard-copy journals and books are archived by putting a thousand copies in a thousand different libraries around the world. Each copy is stable, and the likelihood of all copies disappearing is slight. Electronic media can also have mirror copies, backup copies, disaster backups, and so forth.[14] But each copy is unstable, and sunspots or other natural disasters could wipe out all copies at once. There is still much to learn about the archiving of electronic media. The source [KRA4] explores these questions in some detail.

4.10 Being a Referee

After you have become an established mathematician then you will no doubt be asked to referee papers, or even books. In the case of a paper, there is no monetary reward for the job[15]—you do it as part of your professional activity. For reviewing a book there is usually a modest honorarium.

Almost all refereeing is done blind. That means that the author will not learn your identity as the referee. This gives you license to say whatever you like. But you should exercise that license with dignity and restraint. The right view of the matter is that you are trying to help the author to function as a scholar. Your job as referee is to determine whether this item (a paper

---

[14]One modern commercial publisher has hired three different third-party companies to do backups of its electronic media. The most important single idea in archiving is *redundancy*.

[15]Back in the 1960s there were some Western journals that paid for refereeing jobs. Today a few journals, such as the *Kuwaiti Mathematics Journal*, pay a small honorarium for refereeing.

or a book) ought to be published. Along the way you can help the author with his/her writing, his/her scholarship, and of course his/her mathematics. This should all be done in a constructive manner, just as you would want a referee of one of *your works* to do.

I sometimes fall off the wagon—if I think that an author has been immensely careless or sloppy or irresponsible or just plain stupid. Or if he/she has not adequately cited my own work. I may just lose it, and say a number of uncharitable things. But I always regret it. Most of the time authors are immensely appreciative of my input, and say so (the editor is then good enough to share the sentiment with me, while maintaining my anonymity).

A good referee's report should tell the editor (of the journal or of the publishing house) some of the following things:

- Is this paper/book timely?

- Is it interesting?

- Is it correct?

- Is it written in an attractive manner?

- Are the references complete and accurate?

- Are the graphics well-rendered?

- Is the length suitable?

- Is the result sufficiently important to be published (this query applies mainly to research papers, but the spirit of the question could apply to most any piece of writing)?

- Is this author the right person to have written this piece?

Of course you are free, if you wish, to include a list of errata or typos or errors in English syntax. You may suggest useful references or alternative proofs.

A short paper can have a short referee's report—of a page or so. A longer paper, purporting to prove a truly significant result, probably requires a more detailed analysis, of perhaps several pages. Now the referee must surely consider all the points adumbrated above, but also might address:

- Whether this is the best approach to the result;

- Whether there is a more efficient or more ideal way of proving these theorems;

- Whether the paper takes into account the existing literature and the known approaches;

- Whether the content and substance of the paper justify the length;

- Whether the List of References is careful and complete;

- Whether the paper has a good introduction and a good historical overview, placing this new work into a meaningful context;

- Whether the paper is well-organized and easy to read;

- Whether the paper makes its points clearly, enunciates its results succinctly, and proves the theorems cogently;

- Whether the paper is significant, or the "final word" in a field, or solves an important problem.

Your referee's report should be of a length suitable to the task. A referee's report for a paper can be a couple of pages. That for a book could be longer. You need to say what needs to be said, and make the points that are essential. No more and no less.

And you need to be timely. A referee's report that takes you four years to produce is not going to be all that helpful either to the publisher or to the author. The editor requesting the referee's report will give you an indication of when he/she needs it. If you cannot come close to meeting the deadline, then you should decline the job. Conversely, if you accept the job then strive to meet the deadline. It is really your professional obligation.

4.11 How to Apply for a Grant

In this country the main source of funding for pure mathematical research is the National Science Foundation (NSF). The NSF can give you a grant to support your work during the summer—pay you a summer salary, subsidize your travel, support your graduate students.[16] Or it can provide money to

---

[16]Today the NSF supports fewer than one third of those who apply for grants in pure mathematics. But many fewer mathematicians even apply today (as compared with forty years ago), just because money is so tight and grants are so hard to get.

help you run a lab, employ postdocs, buy computer equipment. The NSF can also subvent a conference that you are running. In addition, the NSF has considerable funds these days to support teaching and educational activities.

Support for mathematics is not limited to the National Science Foundation. The Department of Education has the important GAANN program (Graduate Assistance in Areas of National Need) that provides subsidy to graduate studies for students in subject areas (such as mathematics) that are perceived to be critical to the national welfare. The Defense Advanced Research Projects Agency (DARPA) provides copious funding for defense-related research—much of it mathematical. The Office of Naval Research (ONR) has traditionally been a steady funder of mathematics. The Department of Energy (DOE) has generously funded research in minimal surfaces and mathematical areas that are perceived to be relevant to oil exploration and other energy sources.

You can imagine that each of these funding agencies has its own paradigms and its own needs and its own formalities for applying for funding. I cannot cover them all here. What I shall do instead is to very briefly describe some general principles for applying for a grant with any agency (this could even be a private philanthropic organization—such as the Templeton Foundation). Good general guidelines are:

- Make your writing shine. Your grant proposal will not be reviewed *only* by mathematicians. All sorts of administrators and cross-disciplinary committees and other filtering mechanisms will be looking at your prose. So make it clear, concise, compelling, complete, and accurate.

- Your writing should not be too technical. Of course you want to be precise and on point and to say what must be said. It certainly will not do to say, "I'm going to work on a hard problem. Send some money to help me along." But if your grant proposal is as recondite as a research paper, then even the experts will have trouble slugging their way through it (since those reviewing the proposal will not be in their research-paper-reading mode). Provide more than the usual amount of explanation and background. Give examples. Help to bring the reader up to speed. Bear in mind that not all readers of your proposal will be experts, or even mathematicians.

- Most every granting organization has strict technical rules about how to formulate and format a proposal. This could include **(a)** page margins,

(b) type font, **(c)** length of each section of the proposal, **(d)** topics to be covered, and many others. You must follow these rules strictly and accurately or, likely as not, your proposal will simply be returned unread.

- You must find out what the funding agency is after and you must speak to those needs. The NSF (for instance) is looking for cutting-edge research that will advance the cause of basic science, and will benefit the quality of life in this country. You must describe mathematical research that is current, of broad interest, and will have a real impact. You must be able to describe what your contribution is likely to be *without making it sound as though you have already solved the problem.* You must give careful descriptions of the work that has already been done (by others and by yourself) and how your program will fit into that context. You must give detailed references to the literature. Different funding agencies will have different criteria. The Department of Education is looking at educational goals. The Department of Energy has an agenda that is dedicated to the advancement of energy research. The Templeton Foundation (which does in fact fund mathematical research) is primarily interested in creating a marriage between spirituality and science.

- In the context of the previous bulleted point, you must describe the milieu into which your work fits. There is an ongoing flow of frontline research in your subject area; how does your contribution fit in? How will it advance that flow? You must describe previous results and partial results—both your own and those of other workers in the field. You must make it clear that you know your subject inside out, and what everyone is thinking. You must know the implications of the work you propose to do, and what significant consequences it may have. You must describe in detail your plans for future investigations, because after all that is what this granting agency is supposed to be funding.

- Your proposal must be ambitious and impressive, but it also must be realistic. I once refereed a grant proposal in which the PI (Principal Investigator) proposed to first prove the Riemann hypothesis, then the Poincaré conjecture, then the Hodge conjecture. This from someone who had not written three papers in the past ten years. It rapidly became clear that this particular PI was just having us all on. Of course your proposal will not read like that. The main point is that you want to propose to work on problems, and to achieve results, that (based on your track record) you can reasonably expect to shape and develop.

- You must be punctual. Most every granting agency has strict deadlines for submission of a proposal.[17] If you miss the date, then most likely your proposal will be returned unread.

If you are new to the grant game, then you will find it eminently useful to have a senior academic, one who has (presumably successfully) submitted many grant proposals, look over your draft proposal. If you are lucky perhaps you can get someone who has served on an NSF review panel lately.[18] Listen carefully to his/her suggestions and make whatever changes are needed. You are going to have to exercise some time management skills here, as you must have time to get your draft written, to have friends and colleagues vet it for you, *and also to have the departmental staff help you prepare the budget.* Preparing a grant budget is a highly complex and convoluted activity that involves many levels of university administration. There are cost-sharing and other issues that are way beyond most of us. The main thing that you need to know is that it may take an extra week or more to hammer out the budget for your proposal. *But you still need to meet the advertised deadline.*

A closing thought is this. Universities and colleges like grants. First, faculty who have grants add to the prestige of the institution. Second, many grants kick back a percentage of the bottom line to the host institution.[19] So there is a fiduciary benefit. Third, having a grant is a measure of your scholarly quality. It shows that your work stands up to outside peer review. It looks terrific in your tenure or promotion dossier. So you want to learn how to apply for grants and you want to do it.

---

[17]Sometimes instead of a deadline there is a "target date," and that is more flexible.

[18]The National Science Foundation flies in teams of mathematicians to review all the grant proposals in a particular subject area in any given year. The panel spends two or three days reviewing and discussing the proposals in detail, and then ranking them. The top cut in this ranking gets the grants. The others do not. The last time I served on an NSF panel one of the NSF Program Officers bragged that they had even turned down Fields Medalists. So the standards are tough.

[19]This money is called *Overhead,* or sometimes *Cost Sharing* or *Indirect Costs.*

4.12 How to Give a Talk

Part and parcel of the mathematical research/education life is giving talks. People find talks to be much more accessible than formal research papers. Without a doubt the published research paper is the way we plant our flag, record our ideas for the record, and establish an archive for progress in the subject. But if we want to acquaint our colleagues with the latest breakthroughs in the field, then a fifty-minute talk is a good device for the purpose.[20]

The basic precepts of how to give a good talk are recorded in [KRA2]. We shall not repeat all the details here. Delightful treatments of the topic also appear in [MCC] and [JON]. Many of these guidelines are similar to ones I have already enunciated for writing a good paper. But they take a slightly different form for oral presentation, and are worth repeating:

- Know your audience. Are you addressing the faculty at Harvard, or an annual meeting of the Mathematical Association of America, or perhaps the undergraduates at Bryn Mawr? Each of these putative audiences has different backgrounds, different expectations, and different needs. It is your job to endeavor to meet your audience halfway.

- Know your subject matter. Well, you should. Typically this is material you have created, in which you have immersed yourself for the past year or two. You almost certainly know more about the topic than anyone else in the room. Nonetheless, you should prepare carefully and have everything down cold. Have all the definitions and key results written out succinctly in your notes so that you will not stumble when presenting them to an audience. Have an outline of your main topics so that you can orient yourself during the talk. Have all your sources listed so that you can provide them if requested to do so.

- Be organized. Remember that your job is first to bring your audience up to speed, so that it is prepared to hear what your latest hot new

---

[20]Back in 1941, Saunders Mac Lane was giving the Ziwet Lectures at the University of Michigan on the interface of algebra and topology. At one point during one of the last talks, postdoc Sammy Eilenberg got an alarmed look on his face and bolted from the room. Consulted later, Sammy allowed that Mac Lane had given him a terrific idea, and he suddenly had a vision of new vistas of mathematics to be developed. Eilenberg and Mac Lane ended up having one of the most prolific and fruitful collaborations in mathematical history.

idea is. You cannot just drop it on them cold. You must acquaint them with the terminology and key ideas, bolstered by meaningful examples, so that they are ready to appreciate all the new material that you will present. Introduce all notation and definitions that you will need before you actually use them. Give meaningful—even exciting—examples whenever you can. Draw pictures. The lecture is both an auditory and a visual experience. Take advantage of both.

- Punctuate your talk with milestones. These can be important illustrations, nice examples, memorable results, key ideas, brilliant definitions, or startling applications. The main point is that you don't want your presentation to be a dreary monotone, uninflected by any excitement or insight. Both the *style* of your presentation and the *content* of your presentation should give the audience a sense of a compelling flow of ideas marked by key results and inspiring acuity.

- Practice, practice, practice. When I rehearse a talk today, I just sit in a chair and run through the notes—thinking about various junctures in the presentation and how to make them work. But I have been doing this for a long time. If you are a tyro, then you will want to actually practice in front of an audience—a small group of friends, colleagues, and students. One standard device, if you are scheduled to give an important lecture at an AMS meeting, or at Harvard, or at the University of Paris, is to give a practice lecture to your local seminar. In the latter context, you have a group of trusted colleagues—people who really know you and know your strengths and weaknesses—who will help you along and aid you in crafting the most effective possible presentation.

- Think about the physical nature of your presentation. If you are going to use a blackboard or whiteboard, then learn how to do so effectively. Write so that your thoughts can be read and appreciated. Organize things visually. Do not stand in front of what you write. If instead you are going to use an overhead projector, then learn how to prepare a good and effective overhead slide. Don't put too much on each slide; better to put too little. Learn to pace yourself: It is all too easy to race through your overhead slides. These days it is popular to give math talks using `PowerPoint` or the LaTeX alternative `Beamer` for PC computers or `KeyNote`® for Macintosh computers. This is great, and can be visually stunning, but again you must learn how to use the medium effectively. No matter how well-intentioned you are, if you cannot use your tools properly, then you are going to be a failure as a communicator.

- The use of electronic slides can help you with the pacing of your talk, but this matter will require some thought. Since the slides are prepared in advance, there is a tendency to flip through them too quickly during your presentation. You already know what is on them, so you are not reading them in real time during your talk. But the audience must. And if you are clicking and advancing too rapidly, then nobody will be able to complete their internalization of any of the slides. As a result, the audience will get lost and your talk will be a flop. People need time to ponder each slide a bit as you talk. They will refer to the written words several times as they listen to your remarks. So the pace should be leisurely. About one slide per minute—with not very much written on each slide—is the *most* that you should attempt in your visual presentation.

I have heard talks by good mathematicians in which the mathematician has a *.pdf file of his/her latest paper on the computer and simply flips through it on-screen while talking about it. This is dreadful. Typically the paper is in ten point type, and there are upwards of thirty lines per page. It is nearly impossible to follow such a visual presentation. It is too dense, and there is too much of it on each screen. A typical slide in a math talk should have about six lines, and be in *very large* type. And the lines should be well spaced apart. The Websites

```
http://www.psychology.nottingham.ac.uk/courses
        /modules/statsguides/Effective_Slides.html
```

and

```
http://psyche.uthct.edu/shaun/SBlack/slides.html
```

and

```
http://www.cis.hawaii.edu/GraphicsHome/ServiceBureau
    /SBSlide.html#TIPS
```

give nice discussions of how to create effective slides. The sites

```
http://www.aimath.org/WWN/fourierconvex
    /Folien.Palo.Alto.pdf
```

and

```
http://www.aimath.org/WWN/hilberts10th/demeyer.pdf
```

have some particularly effective slides from some recent math talks.

- If you are going to be using technology in your talk, arrange to check out the room in advance. Some lecture rooms have built-in, elaborate computer facilities with projectors, sound systems, multiple screens, and so forth. It requires a little time and effort to figure out how everything works, and you will likely need some help from the locals. The other option is that a portable computer, a portable projector, and a portable screen will be carried into the room. This will require some setup, and you will need to familiarize yourself with how everything works.

    I once gave a large public lecture—in Australia!—in which I was displaying computer code on the screen and then compiling it in real time to reveal a displayed graphic. I was working in a room with top-notch, state-of-the-art equipment, none of which I understood. I was as nervous as a cat. If the system got hung up in the middle of my talk then everything would be ruined. I spent a good part of the afternoon (the talk was in the evening) checking and rechecking the system, running through the computer routines, making sure everything worked as it should. And I was glad that I did. When I gave the talk I was confident and in control, and everything came off as I had hoped.

- It is rarely optimal, or effective, to actually explain the definitive version of your theorem. Cutting-edge mathematics tends to be rather technical—after all, you have solved a problem posed by experts for other experts; you are answering a question that only the cognoscenti can appreciate fully. When you give a talk, your job is to convey the ideas to a broader audience. So you might *state* the optimal version of your theorem (but I would hesitate even to do that); the explanations will probably be most apposite and most compelling if they are directed to a special case, or a restricted version of the result.

  You certainly cannot present the full proof (if you can, then your result is probably fairly trivial and you should not be giving the talk in the first place). And you should not even consider attempting to do so. Rather, give an indication of what the proof is about. Draw a picture. Indicate the key techniques. Mention a classical technique that inspires your methodology; do an illustrative calculation—if it is interesting and indicative of the big picture; state a pivotal lemma (but don't prove it); discuss a compelling consequence. The point is that your job is to dance around your result and create a sense of understanding with your audience. Your job is *not* to teach them the proof. And you cannot do it, so don't even try.

  Gian-Carlo Rota said that your audience at a colloquium is like a herd of cows. You should only present these heifers with *one* idea, and gently lead them this way or that way. Two ideas will confuse them. Three ideas will create pandemonium. You are like the pied piper, playing his flute and leading them on an entertaining journey. You are *not* teaching them to drink from a fire hose.

- Do not, please do not, run overtime. It's rude, it shows disorganization and lack of professionalism on your part, and it is ineffective. After about fifty minutes, your audience is simply going to shut down. You will be standing up there talking to yourself. What's the point? Your preparation for your talk (which should be considerable) should provide a number of different exit points so that you can gracefully wrap things up after fifty minutes (or whatever is the designated time slot). It is very bad form to, after forty-five minutes, start telling your audience that you don't have enough time, that you won't be able to present all your brilliant ideas, that you should have been allotted more time.

Ridiculous. This is your show, and you are completely in control. Exercise that control by putting on a good exposition that fits into the given time window. All the real pros—from Ed Witten to Enrico Bombieri to Jean-Pierre Serre—always fit their remarks into the given time. And they probably have a lot more to say than you do. So follow the model of our leaders and give a talk that is of the right length. It will be more effective, and certainly better received.

- Have a nice conclusion. Don't just say, "Well, I guess I'm out of time." or "I think I'll stop there." or "That's all, folks." You can do better than that. Draw together the ideas. Review your main points. Reiterate the substance of your main result. Point to some ideas for future work. Give the audience something to take home and mull over (see [EWI] for the thoughts of a real master on this point). It's the least you can do.

4.13   Graduate Teaching

A part of faculty life that is immensely fulfilling and rewarding is graduate student teaching. It is a real plum to teach a graduate course in a subject area of your particular interest. This can mesh nicely with your research program, it can be a source of new ideas, and it can be a way to attract Ph.D. students. An advanced graduate course or seminar can lead to a nice new book or monograph. It can lead to a joint project with some of the students, or with a fellow faculty member who is sitting in on the course.

Teaching a graduate course can be a great deal of work. If the topic is a well-established subject in which there are a number of good texts, then you can teach the course just by following a well-chosen text. If not, then you will be developing the material from scratch—often by reading original papers. In some cases you will be working out the material yourself from first principles. I have taught courses in which I was madly thrashing away proving the theorems for the first time the night before my lecture! Of course the payoff is that you are developing new mathematics *while you are teaching*. This could lead to a new paper, or at least to some new insights that will be valuable in your future explorations.

At Princeton University every tenured faculty member teaches an undergraduate course each semester and an advanced graduate course in his/her research area each semester. This is one of the perks of teaching at a first-class department. At other schools it is a real plum to get to teach an advanced graduate course in your subject area. There are only so many graduate students who are at the right level, your department has many other teaching obligations, and so only so many advanced grad courses can be offered each semester.

It is also the case that most schools have an administration-mandated minimum enrollment in order for a course to stay afloat. If you cannot attract five students, then your course dies (unless you are willing to teach it for love, and receive no teaching credits for your efforts). In any event, there is competition among the faculty for the privilege of teaching an advanced course. Your bread and butter, indeed the department's bread and butter, is teaching calculus, statistics, linear algebra, and the like. That is what you will teach most of the time. The fun stuff is to teach upper division courses and basic qualifying-exam-level course. But the real treat is to teach an advanced course on your pet research topic. You will get to do that every three years or so. And you should make the most of it.

Of course if you, yourself have a lot of Ph.D. students then they *have* to take your course and therefore your course will surely run. If not, then you may have to do some recruiting. At UCLA we actually did some horsetrading in this regard: Faculty member A would go to Faculty member B and say, "I'll make my graduate students take your course if you will make your graduate students take my course." And then everyone would be happy (except, perhaps, for the students).

Another form of graduate teaching is the qualifying exam courses. These are basic real analysis, basic complex analysis, basic abstract algebra, and basic geometry. At some schools we might add to the list numerical analysis, applied mathematics, optimization, algebraic geometry, algebraic topology, partial differential equations, and a number of other topics. These courses are fun because they are real mathematics: You get to state theorems and prove them. You don't have to worry too much about the level or the preparation of your students; you know exactly who the audience is. And they are usually a serious and determined group, because they are preparing for their quals.

One new wrinkle that has been introduced into the teaching of the qualifying-exam-level real analysis course (a large part of which is measure theory) is that, these days, a good many of the students are from economics and finance. This was initiated by the Black-Scholes theory of option pricing. This Nobel-Prize-winning work of Fischer Black and Myron Scholes uses stochastic integrals to create the world's first option pricing scheme. It has really revolutionized the world of finance, and it is *very mathematical*—at a very rigorous level. So you will find finance students in your class struggling to master measure theory. These are good and serious students, but not necessarily as gifted in mathematics as one might like. They will struggle, and you will find yourself doing more hand-holding than is customary. But it is still fun.

4.14   Directing a Ph.D. Thesis

I have written about the directing of a Ph.D. thesis from the point of view of the student in [KRA3]. In this section I shall reexamine the matter from the point of view of the thesis advisor.

Agreeing to direct a student's Ph.D. thesis is a *huge* commitment. There are few other obligations in life that compare. You will be having a close and detailed relationship with this student for a period of three to five years. You will be teaching him/her how to conduct a research life, how to wrestle with a problem, how to deal with frustration and failure, what to do when things aren't working out.

You will meet with the Ph.D. student at least once per week. And more frequently when things are hot or moving. You will be largely responsible for getting this student a job if and when he/she finishes the program and earns the degree. And, if things go well (as they should), then you will have a lifelong relationship with this student. A former Ph.D. student can become a good friend, a collaborator, a confidante, and more.

So you do not agree lightly to direct someone's Ph.D. thesis. You really ought to get to know him/her first. The person might have been a student in one of your advanced classes, and then taken a reading course from you on a more advanced topic. The student might have asked you to be a mentor for some talk or project that is part of the department's graduate requirements. You could have gotten to know the student through a seminar or even on a departmental retreat or field trip.

The point is that you need to have some sense of whether this is a person with whom you are compatible, with whom you can communicate effectively, with whom you are comfortable. Will you look forward to seeing this student each week, or will you want to just crawl under the desk? Will you be proud to tell your friends and colleagues that this person was your Ph.D. student, or will you wish that you had never even met this person?

These are nontrivial considerations, and ones that must be taken seriously. There are some students who bounce around from thesis advisor to thesis advisor, spending a year with this one and eight months with that one. Probably you don't want to be the next in line. At the least, if a student you are unsure of comes and asks you to direct the thesis, then go to the departmental Office and ask to see the student's record. Talk to colleagues about the student. Consult the Vice-Chair for Graduate Studies. Chat up the Chair of the math department.

Directing a Ph.D. thesis can be one of the most rewarding features of academic life. But it can take a lot out of you. Some have argued that directing a Ph.D. thesis is giving up seven papers that one might have written. I don't know how one makes this estimation. I have written new papers *because* of my Ph.D. students, and I have also written very good papers *with* my Ph.D. students. I can't think of any paper or work that I gave up because of a student. I gave of my time, but I've always felt rewarded for it.

The Mathematics Genealogy Project

http://genealogy.math.ndsu.nodak.edu

reports that Roger Temam (among contemporary mathematicians) has had 101 Ph.D. students. That is the record. Andrei Kolmogorov had seventy-nine and David Hilbert had seventy-five. I myself have had sixteen students, and that puts me in a club with just 159 members. Most mathematicians have just a handful of students—five or fewer. Having a student is just so much *work*, and not everyone is up to it. The great Henri Poincaré had only two students. Harish Chandra had none.

If a student comes to you and asks you to direct his/her Ph.D. thesis, take the matter very seriously. Saying "yes" will be one of your more important decisions, and it could be one of the most rewarding ones.

Once a student has signed up with you to write a thesis, then you have to decide how to handle the matter. If you have already directed ten theses, then you probably have a pretty good idea how to get the job done. You know what works for you, you know how to get the student started, and you know how to measure when guidance is needed and how to provide it. If instead you are relatively new at the thesis-advisor game, then you have some important choices to make.

When I first met my thesis advisor, he took me to lunch and we had a friendly chat about this and that. Then he took me back to his office

and sat me down and asked me what I wanted to work on. I told him
about an exciting paper that I had read recently, and that was definitely
the topic on which I wanted to concentrate my efforts. His reaction was
instantaneous: That topic was way too difficult. I would never get anywhere
with it. Of course I was disappointed, but in point of fact he was doing
me a great favor—a favor that only a very experienced and accomplished
mathematician can grant. He could have "given me my head" and let me
go off and struggle with the problem. But that would have been a recipe
for failure, and could have led to me getting depressed and dropping out
of the program. Instead he suggested three alternative problems from three
different parts of our (broadly construed) subject area. All of those problems
were truly substantial as well, but they had the advantage that they were
problems that one could start immediately to chip away at. And that is what
a young graduate student needs.

Many a thesis advisor in fact does not proceed in the fashion just described. Instead the advisor will suggest some reading for the student, with
the anticipation that it will lead to interesting discussions and then thesis
problems will evolve from those discussions. Other thesis advisors will in
effect say to the student, "Go find yourself a thesis problem. When you have
one, come back and give me a little lecture about it."

I find this last approach to be a bit disingenuous. No matter how bright
and willing and able the student is, he/she simply does not have the experience and taste and discretion to pick a good thesis problem. The student
generally has no idea what the hot research areas in mathematics are, nor
what the flow of ideas in the subject is like. The student will have a difficult
time choosing a problem that is simultaneously at the right level, and accessible. Will this be a problem such that, if it doesn't work out as originally
stated, then the problem can be transmogrified into a solvable problem? Will
the solution of this problem require sophisticated tools that are beyond the
range of this student? Will the student have to read a couple of dozen very
difficult papers before he/she can even begin to work on the problem?

This is why we have thesis advisors. A good thesis advisor can get the
student started on a productive path at just the right level, one that meshes
well with the student's interest and background and abilities. If the student
doesn't get the old fire-in-the-guts for the thesis problem, then not much is
going to happen. The problem will not get solved. The thesis will not get
written.

One of the most trying parts of being a thesis advisor is helping the student over the bumps and disappointments that arise when the problem does not get solved in good time. Worse, if the student feels that he/she is not making any progress at all, then what is the advisor supposed to do? There are a number of choices. You could tell the student to go off and work harder. That may have no effect. You can suggest some lines of attack. A very useful and effective approach—if you can pull it off—is to say to the student: "Here is a nice little lemma that you should prove. If you can establish this result, then it will explain the following. Then you can take this next step." Of course the lemma should be something that you are reasonably sure that the student can prove in a few days or a week. This is terrific encouragement, and will help the student to feel that he/she is accomplishing something. And you also will have indicated to the student where that lemma will lead. This is definitely the mark of a good thesis advisor.

It can also happen—and it happened to me when I was a student— that you will have to say to the student, "We don't seem to be making any headway on this problem. Let me suggest something new for you." In my own case, my advisor pulled something out of left field (at least from my point of view). I was genuinely taken aback, as the new problem was in a subject area with which I was quite uncomfortable. But it turned out to be just the right thing for me. I not only solved the problem, but I went on to write 150 papers in the subject area. It has constituted the bulk of my professional life. I can only give my thesis advisor credit for terrific intuition and judgment in picking that problem for me.

A next critical step—once the thesis problem is solved—is getting the student to write up the thesis. You may think that this is the easy part of the process, but nothing could be further from the truth. For the student has most likely never done any serious mathematical writing before, and certainly nothing of this scope and magnitude. This is a huge task, and will require considerable and detailed guidance. Make sure that the student is aware of all the university stipulations about format for the thesis. Find out whether the math department has a TEX style file for theses.[21] Make sure that the student knows TEX. And get the student started.

Begin by having the student write a very detailed outline of the thesis. Have the student make a tentative list of his/her references. Have the student

---

[21] As noted elsewhere in this book, and discussed in some detail in [KRA2], [KRA4], and [KRA6], TEX is the computer typesetting system created twenty-five years ago by Donald Knuth of Stanford University. It has become the default utility for creating mathematical documents.

list what the main results of the thesis will be. Then have the student write a couple of sample sections for you to read. You may find the result of this first iteration to be disappointing. You will have to provide considerable guidance for style and form in mathematical writing. Even though the student has been reading serious mathematics for ten years or more, it is likely that this first draft will not use sufficient enunciations (displayed theorems and lemmas and definitions and so forth), will not provide sufficient examples, will not have material in logical order (statements and proofs of lemmas before they are needed, etc.), and will not have enough detail.[22] The English may be sloppy and inaccurate. The references and sourcing may be incomplete. I have sometimes had to read my students the riot acts when even the third draft was not up to snuff. Frankly, getting the thesis written in a form that you can both be proud of can turn out to be the hardest part of the thesis-writing process. Most people are phobic of writing, and this liability shows vividly at thesis-writing time.

One nasty eventuality that can occur—fortunately not often—is that you will come to the sad conclusion that things just aren't working out between you and the thesis student. This could be because the student is too formal-istic and won't try anything. Or because the student is too sloppy and can't focus on anything. It could be because the two of you have a divergence of styles, or maybe even that you can't get along. I once had a perfectly brilliant student—who is now a successful mathematician, a full Professor at a Group I math department—who came to me one day in tears to tell me that he just couldn't handle the style of geometric analysis. It was too fast and loose for him. He stopped working for me and transferred to another university where he found a thesis advisor who was more to his taste. In short, most anything can lead to a parting of ways. That's life. Do what you can to help the student to land on his feet and find another thesis advisor. I am assuming here (of course) that you have played the game properly up to this point, and you are simply faced with an untenable situation that needs to be dealt with. You both need to chart a new course and move on.

I can assure you that the majority of advisor-student relationships come to fruition and are a happy experience for everyone. They lead to positive long-term relationships and often to new collaborations and new discoveries. The occasional bumps and lumps are part of the game, and you will learn to

---

[22]In my own students' theses, I usually insist that nothing be left to the imagination. *All* details must be provided.

deal with them.

There is an art to being a good thesis advisor. Some advisors, in an effort to give the student a problem that is accessible and doable, end up giving the student something that is really not very interesting, and that is also a mathematical dead end. The first attribute will make it difficult for the student to get a good job; the second attribute means that the student will leave graduate school not having a viable research program. He/she will not know what to do when it comes time to establish a scholarly reputation—to write papers and to publish. Of course if you, yourself go to seminars and conferences and keep in touch with what is going on and talk to people regularly, then you will be well-informed and you will know all the worthwhile problems. That will make it more likely that you will have plenty of good problems for students. At one point the eminent Marshall Stone said that he wasn't directing any more theses because, "If I knew any good problems I would work on them myself." One can only suppose that this was said in jest; but you want to be sure that you are never caught out saying such a thing yourself.

4.15 Professional Travel

A mathematical colleague of mine once remarked that, even though he was not wealthy, he felt as though he were wealthy. What this meant was that he got to travel to far-flung parts of the world and meet all sorts of interesting people and see exotic places.

Certainly this is one of the notable perks of academic life. Especially if you publish, and if your work is appreciated, you will get invitations from all sorts of places—to spend a few days, or a week, or a month, or perhaps a semester or a year. You could even get offered a permanent job, but let me save that particular topic for another part of this book.

Usually an invitation of this sort includes a promise to pay your travel expenses—hotel and airfare at the least. It could also include a *per diem* (to cover meals and incidentals). And it could include an honorarium. The bottom line is that, basically, you get to travel for free. And the price you pay is that you must give one or more talks and you must hang out and talk with the other mathematicians. Not a bad gig.

Some invitations—such as an invitation to give a talk in a special session at an AMS meeting—will include *no offer of subvention* whatever. Just because there are no funds for these activities. The hope is that you will have an NSF grant that will help to cover your expenses, or perhaps your department will kick in some funds. NSF grants are getting harder and harder to obtain: there are many more topnotch people applying and—what with all the federally funded math institutes and all the new NSF programs—fewer funds. Departments are also strapped for money, but there is almost always some sort of fund to help defray travel costs for faculty. Chances are that your department will be delighted to know that you've received a nice invitation; it will bend over backwards to find some funds for you.

There are many more paradigms for funding your travel than the above discussion reveals. For instance, if you are invited to a conference at the International Research Center at Banff, you will be offered free room and board during your stay. But you must find your own funds for the plane fare. From Banff's point of view, this setup makes perfect sense. For their funds are limited. Since they pay only room and board for each participant, they know in advance exactly how much each participant will cost them. They don't have to deal with some poor scholar from Siberia who has a $3000 plane ticket. If you are invited to spend time at the Mathematical Sciences Research Institute (MSRI) in Berkeley, you will be offered some fixed stipend. For example, you might be invited to spend a semester and be offered, say, $10,000. Of course that amount of money will not cover your costs for five months, but it is seed money and the hope is that the rest of your expenses will be covered by sabbatical funds or a grant.[23]

One unusual setup is the American Institute of Mathematics (AIM) in Palo Alto, California. When you are invited to a one-week workshop at AIM, then all your travel, all your hotel, plus a per diem are covered. There is no other math institute in the world that can do this. AIM is funded in part by the National Science Foundation and in part by Fry's Electronics.

Sometimes you will have to spend some of your own resources to cover your professional travel. Usually you will not. One nice feature of this type of travel is that you can sometimes coordinate it with a personal trip. Perhaps your spouse can join you at the end of the conference and you can spend a week sightseeing (this option is probably more attractive in Paris than in Detroit). Or your spouse can spend a week hanging out with you at the

---

[23]An exception to what was just said about MSRI is that there are *full* stipends for beginning mathematicians. These are actually a salary, and cover *all* expenses. Such a position is in effect a postdoc.

conference—entertaining himself/herself during the day and socializing in the evenings—and then you will have a week to yourselves. The professional part of the trip is still tax deductible and/or reimbursable; the personal part of the trip is of course "on you."

4.16   Sabbatical Leaves

The word "sabbatical" has both Greek and Hebrew antecedents. It was a custom among ancient Jewish farmers to take every seventh year off. Thus it has come about that (at least in theory) every seventh year the professor gets a year off.

The purpose of the sabbatical is to give the professor some time to recharge his/her batteries, to learn new things, to write a book or conduct a collaboration. It is a way to help him/her keep his/her scholarly program alive.

Well, like all good things in this life, the interpretation of what "sabbatical" means has evolved, and it is going to vary considerably from school to school.

At the University of California the situation is quite structured. For each six terms that you teach you get a term off. Period. You don't have to compete for it or justify it; it's yours. You are given full salary during the time off, and you can spend it building a house if you like (in fact I know one professor at a U.C. campus who did just this, and he gave a student an Independent Study Course to help him).

At my own institution, after you give full service for six years then you are entitled to a sabbatical. You can have one semester at full pay or a full year at half pay. You don't have to compete for this largesse; it is your right.

At a great many schools, especially state schools, you have to compete for your sabbatical. The Dean only has so many to give out, and the number is considerably fewer than the usual "rule of seven" would dictate. So you have to complete some paperwork in order to convince the Dean that you deserve a sabbatical. This may include reminding the Dean of all the university service you have been performing. And it will certainly entail making a convincing case that you are going off for a year or a semester to do something worthwhile. For example, when I was at Penn State, it was a surefire hit with the Dean if you told him you had arranged to spend a year at Harvard to sit at the feet of some Nobel Laureate to learn the latest ideas in string theory (or pick your favorite hot area of scholarly activity); if instead you told the Dean that you wanted to spend the year at the University of Southeast Idaho to write a calculus textbook with your former Ph.D. student, then he was less likely to be moved.

Let me stress that sabbaticals are not available to adjuncts and Instructors/Postdocs. The sabbatical is, along with tenure, one of the great perks of being a tenure-track academic. You don't need tenure (i.e., you don't have to be an Associate or full Professor) tosabbatical!privilege of tenure-track faculty take a sabbatical. You just have to have punched the time clock for the right amount of time.[24] The sabbatical rules at your institution are public knowledge; be sure to apprise yourself of the law of the land, and to avail yourself of this privilege whenever you can.

A final note is this. Faculty also have opportunities to take "leaves of absence". How does a leave differ from a sabbatical? Usually you take a leave when some other institution offers to pay your full—or at least a substantial amount of your—salary and benefits for the year or the semester. This makes it easy for your Dean to let you go away for a time—because it doesn't cost the Dean anything. The Dean will start to get nervous if you ask for a second year's leave (because he will worry that you are not coming back). Most Deans will absolutely forbid a third consecutive year of leave.

You can imagine that there are certain faculty who have manifold opportunities to take leaves: If you are a world-class scholar, recent winner of the Fields Medal or the Wolf Prize, then of course every institution would like to have you around for a while. If you do applied mathematics, then perhaps NASA or some other government agency would like to have you as a paid consultant. Maybe some big company—like Hughes Aircraft—would like to put you on staff for a year. It is easy to see how faculty may find these opportunities attractive, and it is also easy to see why the Dean wants to keep them under control.

---

[24]Some Deans, like my own, are willing to be quite generous with sabbaticals for Assistant Professors. Of course if an Assistant Professor waits until his/her seventh year to apply for a sabbatical, then he/she will already be tenured (or not). Our Dean, and some other Deans, feels that a young scholar deserves some time without duties to develop a strong research program. So he is willing to give a sabbatical to an Assistant Professor with only three or four years in the saddle.

4.17 What Clique Do I Fit Into and How Is It Regarded?

The world population of mathematicians is rather large—perhaps 60,000 or more. But you will only interact professionally with a fraction of these.[25] In particular, in your research life, you will have a specific group of people with like interests and with whom you regularly communicate and share ideas.

It is a simple fact of life that you will be judged—in letters of recommendation, and for tenure, and for promotion—in part by the crowd that you hang out with. If you are an analytic number theorist working on the Riemann hypothesis, then it is safe to say that you are front and center in the modern mathematical enterprise. The leaders of that gang are winners of the Cole Prize, the Fields Medal, and other major encomia. It is a large and active group with many seminal achievements to its credit. So if you are part of this program, and if a letter about you says, "This is one of the three best junior people in the subject," then that really says something—and it says something that *any mathematician* can recognize and appreciate.

If instead you work on Moufang quadrangles, or non-Euclidean elliptic geometries, or Gelfand-Fuks-Virasoro cocycles, then you will be a member of a rather exclusive club. There may be only ten people worldwide in your entire field. So if a letter about you says that "This is one of the three best ...," then that may be only out of a group of four. Not quite so impressive.

I am not trying to be judgmental here. Moufang quadrangles and the other areas mentioned are perfectly useful and interesting parts of the mathematical landscape. But they are tightly focused areas with just a few participants. If you are in one of these fields then someone reading letters about you, or evaluating you for a position, or assessing you as a potential colleague will have a difficult time telling who you are and what you do and with whom (that they have heard of or will know) you might be compared. It is a little bit like being a violinist. If you are being assessed in the context of your performance with the Boston Symphony, then one picture will emerge. If instead you are being assessed as the First Violinist of the Anaheim String Band, then a different picture will take shape.

I once witnessed a rather draconian personnel situation at a major re-

---

[25]It is a bit like living in the United States. The country has a population of 300 million, but you live in a state, and in a town, and in a community. You probably only actually *deal* with a few hundred people.

search university in California. A tenure candidate had been hired to be a seed to their applied math program, but he had switched fields—under the tutelage of a senior mentor—to a branch of discrete mathematics. At his tenure "hearing" in the department, this fellow was unmercifully crucified for **(i)** switching away from applied mathematics (and thus hurting the department's expectations for developing its program) and **(ii)** moving into a field that was "obviously" third-rate. This in the face of his senior mentor, a scion of that third-rate field, being front and center at that very meeting. It all went very badly, with a good many hurt feelings, and the candidate was brutally defeated for tenure. I am happy to say that he landed on his feet and got a good job elsewhere. But this situation illustrates the idea that not all fields are equal, and one can be judged not just on individual merits (in fact this particular candidate was quite good in his new field) but also on the position of one's work in the overall tapestry of mathematics. Put in starker terms: One can be judged for which clique one hangs out in.

In today's electronic world there are no secrets. Anyone at all can get on *MathSciNet* and do a search for papers whose titles contain the same key words as do yours. They will thus be able to tell quickly how much activity there is in the area, and who is engaged in that activity. Is it Chair Professors at Harvard, or is it people of lesser stature? Another Web search will tell immediately how many conferences there are in the subject area, who are the organizers, and who the principal speakers. In about five minutes someone assessing your standing in the subject area will have a fairly accurate idea of what context you fit into, how important that context is, and where you stand in the great scheme of things.

Of course we all must grow where we are planted. Your subject area will be largely determined by who your thesis advisor is and what sort of problem he/she gave you. If you are a real knock-me-dead mathematician then, once you are out on your own you will chart your own course and move into various fields that strike your fancy. You will not be tied to the program that your advisor gave you. Most people, however, stick to familiar ground. Ten years after the thesis they are still writing papers in the general area in which they began. That is fine; but that is the world you will inhabit and you will be judged by that standard.

4.18 The Mathematics Research Institutes

Today, in the United States, there are seven federally funded mathematics Institutes. These include

- The Mathematical Sciences Research Institute in Berkeley

- The American Institute of Mathematics in Palo Alto

- The Math Biology Institute in Columbus

- The Institute for Mathematics and its Applications in Minneapolis

- The Institute for Advanced Study in Princeton

- The Institute for Pure and Applied Mathematics at UCLA

- The Statistics and Applied Mathematical Sciences Institute at Research Triangle Park in North Carolina

And there are yet other institutes. The Clay Mathematics Institute in Cambridge, Massachusetts is funded entirely by private money from venture capitalist Landon Clay. The University of Maryland math department has the Norbert Wiener Center for Harmonic Analysis and Applications. In Canada there is the Math Research Station at Banff and the Fields Institute in Toronto. There is the Pacific Institute of Mathematical Sciences. There are a good many research institutes in Europe and Asia and other parts of the world. Compared to 150 years ago, we have almost an embarrassment of wealth when it comes to facilities to support mathematical research.

What good do these institutes do for the average mathematician? First, any of them is a great place to spend a sabbatical leave. You will be immersed in a stimulating research atmosphere, away from the daily vagaries of departmental life, be supported by a very professional and energetic staff, and surrounded by your colleagues. What more could you want?

It is also possible for you to *organize* an event at one of the Math Institutes. This could range from a one-week conference or workshop to a semester-long program. If you go to the Web page of any of the institutes it will tell you the procedures for setting up a program and obtaining funding. Of course it is not automatic. You apply to run a program and the institute's Scientific Board will review your proposal and make a decision. If the decision is positive, then all the resources and support mechanisms of the institution will be put at your disposal. They will help with the invitations, with finding housing for the participants, with providing facilities, and so forth. There will also be money provided to help subvene the travel and living costs of the participants. This will generally *not* be full funding; partial funding will be offered, and it is hoped that sabbatical support or perhaps grant funds can provide the rest.

Even if you are not a registered member of a formal event at one of these math institutes, most have an open-door policy. That means that, on your own funds, you can drop in to hear lectures or participate in discussions or often even take part in social events. Mathematics is a free and open community and we welcome everyone to our activities.

4.19   Outside Offers

If you are a bright and active mathematician working in an area of mathematical research that is particularly hot, then your professional reputation will be well and widely known. Even though mathematics has grown considerably in the past forty years, the discipline remains highly connected. Everybody has a sense of who everybody else is. Your talents will be valued not just at your home institution but across the profession.

Well, the truth of the matter is that your home institution can easily become jaded. Once they've got you and given you tenure and given you an office and a salary, they stop worrying about you. Why would anybody leave? This is, after all, such a wonderful place.

Perhaps it is, but the grass can be greener in a variety of different ways. You could be a faculty member at University X and then University Y— with a working group of five people in your subject area—may decide to make you an offer. Wouldn't it be great to have five colleagues with whom to share your mathematical ideas on a daily basis? Perhaps University Y is an elite institution that attracts top-notch students. Maybe University Y has particularly nice salaries. Maybe you just like the location of University Y—an especially attractive city or a nice college town. It is possible that University Y is an especially wealthy institution that can offer all sorts of research support and creature comforts that your home institution cannot.

To make a long story short, you may get outside offers. This will give your Chair a headache. If another institution values you, then it is quite likely that your home—University X—does not want to lose you. And University

X is not accustomed to putting forth any effort to hang on to you, so now it is going to have to get its act together. It will have to come up with a competitive salary—to meet or match your outside offer. And it will have to come up with perks that will make you just about as happy as the new job would have.

The home institution will not have to try as hard as the new invader, because if you stay where you are then you won't have to move, you won't have to uproot your family, you won't have to get involved with all the paperwork and other details connected with changing venues. So your counteroffer from University X may be a bit lower in a number of respects than the outside offer. And you will have to determine how to put this difference into perspective.

The invading institution will have you in for one or more visits, and your spouse as well. They will make special efforts—much more probably than your home institution has done—to do nice things for your partner (find him/her a job, set up interviews, arrange for him/her to meet people). They will arrange for a realtor to show you houses and condominiums. They will acquaint you with the schools for your children. They will take you out for nice meals. In short, they will make every effort to turn your head.

As I have said elsewhere in this book, it is healthy to reinvent yourself and do something new every five or ten years. So you may find it to be just the ticket to take a new job now and then. After all, you are not going to get this kind of attention for your entire life. When you do get it, you may as well enjoy it and try to derive as much from it as you can.

Although outside offers are most commonly given for research qualifications, there are other reasons that they happen. You could be the progenitor of an important national teaching program. You could have created a new mathematics curriculum. You could have been a leader in the teaching reform movement. You could be an applied mathematician who can attract substantial funding for big programs. You could be famous for getting big grants to alter the landscape of the teaching profession. You may have proved yourself to be a gifted Chairperson who can truly make a difference in the quality of life of a department. You could be a potential Dean. Even in late middle age, people can get offers to relocate because of qualifications such as these. Many things are possible.

There are a number of things to consider when you are entertaining an outside offer. First, you should consult your spouse or significant other to see what their point of view may be. Will this be an improvement? A welcome change? Or instead will it lower the overall quality of life? Another—although perhaps imponderable—consideration is: Will this be the last outside offer you will ever get? Is this your one and only chance to make a move? Or should you wait for something better? Finally, do you owe something to your current department? Will your departure severely affect its programs, its research profile, or its national ranking? Do you care? If so, why? You should have some frank conversations with the department Chair, and go over these points as appropriate. When I have had outside offers, I not only met with the Chair but also with the Dean. They both wanted to know why I was considering the outside offer, and what particular features I found attractive. And also what I expected the home institution to counteroffer.

You are, after all, part of a team. If you are considering joining another team then there are many factors at play. You should consider them all with appropriate weight.

4.20   What Goes On at Conferences?

In the seventeenth century, and on into the eighteenth century, there were few mathematics conferences. After all, there were few professional mathematicians. Recall that Isaac Barrow had to give up his post at Cambridge so that Isaac Newton could have a Professorship. A titan like Bernhard Riemann did not obtain a Professorship until he was lying on his death bed. Galois and Abel never had Professorships. Fermat was strictly an amateur mathematician—his "day job" was as a Judge in Toulouse.

And there was no money to finance mathematics conferences in those days. There was no National Science Foundation, no Office of Naval Research, no DARPA, no UNESCO. Further, there was a certain inbred cultural bias against open communication. The very first scientific journal was founded by Henry Oldenburg in 1665. At that time there was a standing tradition for scientists to be secretive about their work. Oldenburg had established himself as something of a go-between among various eminent scientists, arranging to trade scientific insights for valuable bagatelles like books. His *ad hoc* marketplace of ideas ultimately evolved into *The Philosophical Transactions of the Royal Society of London.*

But the proclivity for secrecy among scientific researchers did not instantly disappear with the advent of Oldenburg's journal. Even two hundred years ago, scientists and mathematicians tended to work in isolation. Because there were so few professional positions, people tended to harbor strong jealousies. Some mathematicians had wealthy patrons (for example Euler had Frederick the Great), but those were the lucky few.

So, while the resources to fund conferences were not available in the old days, it is also the case that the desire for conferences was at a low ebb. The International Congresses of Mathematicians, which began around 1893, were something of a revelation. They ushered in a new era of international communication among the top workers in our subject.

Today we are in a golden age of mathematics, and mathematical communication. Wonderful new cross-fertilizations are taking place among widely disparate fields of mathematics. Fruitful new symbioses are arising among mathematics, engineering, physics, and many other parts of modern technology. And this activity is all spurred by an explosion of communication. For the most part this exchange of information is wide open. Few of us are secretive, and there is little motivation to try to be so. There are myriad sources of funding for conferences—from individual universities to government-level scientific foundations (the NSF being a prime example) to branches of the armed services to private foundations. Most active research mathematicians go to at least one conference per year, and many go to several. It is a great pleasure to organize a conference—to gather a collection of your comrades-in-arms and spend a week exchanging ideas and insights and problems. Also to drink some beers and have some good meals together and to go on hikes or to museums.

So what goes on at a conference? There are many different formats. The most basic conference will have a menu of talks all day. If it is well-organized, then there will be plenty of breaks and a long lunch hour so that people have a chance to have private conversations, to exchange ideas, and to discuss the talks. Some conferences will have plenary, one-hour talks in the mornings and then a lot of shorter talks in the afternoons. Many times the shorter talks will run in parallel sessions so that everyone who wants to will get a chance to speak. Sometimes there will be special talks to help bring the graduate students up to speed. Or there will be a problem session, often chaired by an especially distinguished member of the group.

It is quite common at a conference (running Monday to Friday) for Wednesday afternoon to be free. That way people can go sightseeing, or get together with collaborators and work on problems, or go on a hike. Of course evenings are usually free, and people can go out to sample local cuisine, to the theater, to an opera, or to some other cultural event.

The American Institute of Mathematics (AIM) in Palo Alto, California has for the past five years pioneered a new type of mathematics workshop in which there are two expository talks each morning followed by a breakup into working groups in the afternoon. Each working group, of size about five to eight, tackles some very specific mathematical problem or group of problems. This has turned out to be a daring and highly productive way to get people together to produce new mathematics.

I have been going to conferences for thirty-five years now. When I first started out, I made a point of going to as many talks as I could. That was probably appropriate, as I was trying to tool up; I wanted to absorb as much mathematics as possible. Now my point of view is quite different. I go to conferences to learn new problems to work on, and to establish new collaborations. For me, going to talks is an activity of definitely secondary interest. I want first to meet people and exchange ideas.

You will develop your own style for participating in conferences. I know people who will go to a conference and spend most of the time sitting in their private rooms thinking about some new idea. Others will spend the whole time talking to a collaborator and not attend a single talk or event. Still others will attend several talks but will not pay any attention; they will instead have side conversations (during the talk!) with their neighbors.[26]

The main point of a conference is to get your creative juices flowing. It is important, if one wants to be a successful mathematician, to be immersed in the current activity, to be exposed to all the new ideas. A conference is an effective vehicle for making that happen.

You may find it attractive at some point to organize a conference. You may wonder how this is done. In point of fact, the American Mathematical Society (AMS) can make this easy for you. You can organize a special session at an AMS meeting. This is a gathering of twenty or so experts who each give talks, over the span of a couple of days, dedicated to a particular subject area. Much of the infrastructure—the hotel accommodations and the rooms for the talks—will be done for you by the AMS. All you need to do is provide the theme and select the speakers. Unfortunately, there are usually no funds for a special session (many times a participant's department, or an individual NSF grant, can help out). But organizing such an event can give you a small-scale introduction to the activity of running a conference.

You may also be interested in organizing a full-blown conference of one

---

[26]Einstein and Lefschetz were famous for doing this.

or more weeks' duration. Of course you will not do this math autonomously. The standard device is to form a committee of about four or five people, apply for funding (to one or more universities, to the National Science Foundation, to the Humboldt Foundation, or to some other generous source), and (in concert with your committee) put together an invitation list. The fun part is that you get to shape the event, to specify the activities, and to choose the speakers. You can have a significant impact on your subject by organizing just the right event at just the right time. If the conference is successful, you may wish to publish the proceedings. Again, such a volume can prove to be important for your subject area, and being the editor of such an effort can be a feather in your cap. Certainly organizing a conference is a significant way to contribute to the profession, and also to develop working relationships with your colleagues. It is a worthwhile and rewarding thing to do.

4.21   The International Congress of Mathematicians

The International Congress of Mathematicians (or ICM) is a watershed event in mathematical life. It has taken place every four years for the past hundred years or more (with hiatuses during the great wars). Ever since 1936, the Fields Medals have been awarded at the ICM. Although few of us harbor any illusions about winning the Fields Medal, we all look to this ceremony as a seminal event that helps to define who we are, what is important in our field, and where we are going.

The ICM is *not* like a working mathematics conference. You don't go there expecting to hook up with a potential collaborator and to start writing a new paper. It is largely a ceremonial event.

There are three kinds of talks at the ICM. The plenary lectures—of which there are about fifteen—are one-hour talks given by the most distinguished and accomplished mathematicians in the world. It is a tremendous honor to be asked to give a plenary talk at the ICM—some say a greater honor than to win the Fields Medal. These talks are generally of very high quality, and well worth hearing. The second type of talk is the "invited talk". These are each forty-five minutes in duration. They are given by those who are considered to be the current movers and shakers in the various fields of mathematics. There are several dozen of these talks. These talks are to be taken quite seriously, and they are generally good. But they are often quite technical, and you cannot expect to understand the invited talks outside of your special field. Finally, there are the "uninvited talks". These are volunteer talks; a participant will submit an abstract, and then he/she can give a brief talk (usually fifteen minutes or less). Often a mathematician will give one of these talks to validate his/her participation in the conference so that he/she can receive remuneration from his/her math department. These talks are often forgettable, although there are some notable exceptions.

The main purpose of going to the ICM is to soak up some mathematics, to get a sense of what is going on in our discipline, and to meet people. There are lots of ancillary social events as well. The 2006 ICM was special because people were brimming with excitement to learn about the status of the Poincaré conjecture. Richard Hamilton and John Morgan gave splendid talks explaining the mathematics in broad and accessible language, and assuring the audience that the theorem really had been proved. This atop the awarding of the Fields Medal to Grigori Perelman, and the fact that he declined the honor and did not attend the ceremony.

I would recommend that you go to at least one ICM in your lifetime. It is a good excuse for some foreign travel, it is lots of fun, and you will be exposed to the political/social side of the profession in new ways. Organize a group from your university—that will make it even more fun. Bring the family. Make it into an outing.

4.22   The January Joint Mathematics Meetings

The American Mathematical Society (AMS) and the Mathematical Association of America (MAA) jointly sponsor a big math meeting each year in January. The confab lasts about four days, although there are a number of preliminary activities (seminars and convocations of one sort or another) that can extend the meeting to a week or more. There are several features that make this an attractive event to attend:

- The meeting is very well attended. Well, better attended if it is in San Francisco than if it is in Cleveland. But there are always several thousand people. And it attracts a broad cross section of the profession, from junior college teachers to liberal arts college teachers to comprehensive university faculty to university professors. And a good many foreign visitors attend as well!

- You are liable to see lots of friends, collaborators, colleagues, and old acquaintances. And you will meet a lot of new people in addition.

- There are a number of important invited lectures, and these are generally quite good and very informative. Many of these are big events, and have receptions and dinners connected with them.

- There are quite a number of working seminars or "special sessions" in particular areas of mathematics. These are gatherings of fifty or more people in a particular research area, with a number of brief (twenty-minute or so) invited talks. A special session could comprise three separate three-hour events stretching over three successive days.

- In the past ten or fifteen years there has additionally been a significant increase in the amount of mathematical sociology at play at these meetings. I have seen special sessions on feminist linear algebra, on the role of gays in the mathematics profession, on sexual harassment in the classroom (including instructive skits), and on how to get a grant.

- There are a number of panels about issues of current interest. This could include helping young people to get jobs or teaching with electronic media or techniques for applying for grants or the role of the mathematician in society. Usually the panelists are distinguished people who have something significant to say. A good panel will generate considerable audience participation. Panels can be stimulating and informative and I recommend that you sample some.

- There is a huge exhibit area for publishers and others who market mathematical artifacts. This will include all the mainstream mathematics publishers, the textbook publishers, the calculator purveyors, the software creators (TeX,® Mathematica,® Maple,® MatLab,®and the like), and also various instructional programs (the MASS program at Penn State, various summer programs for high school students, the R. L. Moore project, the Mathematics Genealogy Project, and so forth). Various math institutes will have booths. The National Science Foundation (NSF), the National Security Agency (NSA), and other government agencies will have booths. Various artists and artisans who produce "mathematical" artwork and jewelry will have exhibits. One can spend a good deal of time with these folks, and one should. It is a big slice of the mathematical pie. If you are thinking of publishing a book, then this is a good place to meet up with a publisher.

- Many of your colleagues and acquaintances and pals will be at the meeting. You will have a chance to hook up with old friends and maybe even to talk good mathematics. Or at least to exchange some gossip and reminisce about mutual acquaintances.

- A number of nice outings are offered to see the local sights.

- There are a number of banquets and receptions and other social events.

Of course there are expenses connected with attending the national meetings. Apart from travel and hotel, there is a registration fee of about $250 (considerably less for students and the unemployed and the retired). If you don't register and thus don't have an approved name tag, then you will not be able to get into a lot of the events. Of course it is all tax deductible, and you can likely get your department to help with some of the expenses. The American Math Society also has some grants to help people attend the event, and their Website explains how to apply for these.

Often, but not always, the main meeting is in the host city's convention center, and the canonical lodging is a hotel connected to the convention center. Sometimes the meeting will be in a couple of hotels with no convention center involved. Sometimes some of the events are at a local college or university. It all depends on the physical layout of the community. Generally all the activities will be within walking distance of the hotels.

Your math department will probably at least partially subvent your participation in the joint national meetings, especially if you are a participant (either giving a talk, or serving on a panel, or organizing an activity). You can learn more about the joint meetings at `http://www.ams.org/meetings`. On that Website, you can register for the meeting and arrange your hotel accommodations. You can also sign up for various seminars and excursions and banquets. These meetings are a big mathematical social event, and many of us go religiously every year. It is a way to keep tabs on the profession, and to stay plugged in to current developments. Consider becoming part of it.

4.23 Prizes and Encomia

A subject like chemistry is littered with prizes—encomia to recognize outstanding research or service or teaching or excellence in some aspect of the professional life. Nearly every practicing chemist has some prize or other.

Mathematics, by contrast, has traditionally been rather chary with prizes. The last few years have seen the emergence of several new ones, but our aggregate number of prizes is still relatively small. I mention just some of them here (in no particular order):

The Fields Medal This is the pinnacle of all math prizes. Instituted by John Charles Fields in the 1920s, this honor has become the preeminent mathematical encomium. Sometimes called the "Nobel Prize of mathematics" (there is no formal Nobel in mathematics), this honor is bestowed every four years at the International Congress of Mathematicians on between two and four young mathematicians under the age of forty.

The Leroy P. Steele Prizes Given by the American Mathematical Society each year, there are three of these awards. One is for a seminal paper, one is for an entire oeuvre of scientific work, and one is for quality writing.

The Bergman Prize This prize, founded in 1988, is funded by the estate of complex analyst Stefan Bergman. Administered by the American Mathematical Society, it recognizes outstanding work in fields that were of interest to Bergman (complex analysis, several complex variables).

The Birkhoff Prize Named in honor of George David Birkhoff, this prize was founded in 1967 to recognize excellent work in applied mathematics.

The Moore Prize Named in honor of E. H. Moore, this prize was established in 2002 to recognize an outstanding research article in one of the AMS primary research journals.

The Robbins Prize Named in honor of David P. Robbins, this prize was established in 2005 to recognize work in algebra, combinatorics, or discrete math that has a significant experimental component.

The Wiener Prize Named in honor of Norbert Wiener, this prize was established in 1967 to recognize outstanding work in applied mathematics.

The Conant Prize Named in honor of Levi L. Conant, this prize was established in 2000 to recognize an excellent expository paper in either the *Bulletin of the AMS* or the *Notices of the AMS*.

The Morgan Prize Named in honor of Frank and Brennie Morgan, this prize was established in 1995 to recognize excellent research by an undergraduate.

The Veblen Prize Named in honor of Oswald Veblen, this prize was established in 1961 to recognize outstanding work in geometry.

The Abel Prize Created in 2002 by the Norwegian government, this prize is modeled after the Swedish Nobel Prize and is intended to recognize excellent work in mathematics.

The Satter Prize Named in honor of Ruth Lyttle Satter, this prize was founded in 1990 to recognize outstanding mathematical research by a woman.

The Wolf Prize Funded by the Wolf Foundation and awarded in Israel, this prize recognizes scientists and artists in a variety of fields. Among these are mathematics and physics.

The MacArthur Prize Funded by the John and Catherine MacArthur Foundation, this prize recognizes excellence in all fields of human endeavor. A number of mathematicians have been so honored, most recently Fields Medalist Terence Tao. The premise of the prize is that extremely talented individuals should not be burdened with worrying about worldly peccadillos like paying the bills. The cash value of the prize is five years' salary.

The Cole Prize Named after Frank N. Cole, this is one of the preeminent prizes in algebra and number theory. Founded in 1928, the prize is administered by the American Mathematical Society.

The Bôcher Prize Named after Maxime Bôcher, this is one of the top honors in mathematical analysis. It is administered by the American

Mathematical Society. The oldest of the AMS prizes, this one was founded in 1923.

The Ostrowski Prize A mathematics award given every other year by an international jury from the universities of Basel, Jerusalem, Waterloo and the academies of Denmark and the Netherlands. The prize, founded in 1989, is funded by the estate of Alexander Ostrowski.

The AMS Book Prize—recently renamed the Doob Prize This is a relatively new prize, first awarded in 2005 and given every three years thereafter. Administered by the AMS, it is for an outstanding book that has been profoundly influential.

The Euler Prize Named after Leonhard Euler, this prize was endowed by Paul and Virginia Halmos to recognize an outstanding book about mathematics. The prize was first awarded in 2007.

The Chauvenet Prize Named after William Chauvenet of Washington University, this is a prize usually given for an outstanding expository article—often one that has appeared in the *American Mathematical Monthly*. Founded in 1925, the prize is administered by the Mathematical Association of America.

The Lester R. Ford Award The Lester R. Ford Awards were established in 1964 to recognize authors of articles of expository excellence published in the *American Mathematical Monthly* or *Mathematics Magazine*. The award is administered by the Mathematical Association of America.

The Allendoerfer Award Beginning in 1976, this prize was created for papers in *Mathematics Magazine*. The prize is administered by the Mathematical Association of America.

The Crafoord Prize Founded in 1980 by wealthy Swedish industrialist Holger Crafoord and his wife Anna-Greta, this prize (somewhat analogous to the Nobel Prize) is awarded in fields complementary to those recognized by the Nobel Prize: astronomy, mathematics, geosciences, biosciences, and polyarthritis (a disease which afflicted Crafoord late in his life). Like the Nobel Prize, the Crafoord Prize is awarded by the Swedish Academy.

The Franklin and Deborah Tepper Haimo Award This is a national award for mathematics teaching, administered by the Mathematical Association of America. Founded in 1993, the prize is named after Frank and Deborah Haimo, both outstanding St. Louis mathematicians.

A Plenary Talk at the ICM This is not a prize as such, but it is better than most prizes. Because it plants you firmly in the pantheon. Only the most esteemed and accomplished mathematicians in the world are asked to give a one-hour talk at the ICM. Everyone will remember that you were a plenary speaker. It is a very big honor. There is only one classical mathematician (Volterra) who has given four plenary talks at the ICM. Ahlfors is the only one who has given three. And just a handful have given two. Even to give one is a huge deal.

For most—but not all—mathematics prizes, the cash value of the prize is of little consequence. It is a considerable honor to garner one of these awards, and the concomitant reputation will follow you around for your entire life. Whenever you are introduced—as a colloquium speaker or honored visitor— it will always be mentioned that you are a holder of the Chauvenet Prize, or the Steele Prize, or the Fields Medal. Most of us do mathematics for the love of the game, and not in pursuit of prizes. But it is a special honor and pleasure—and a privilege—to have one's work singled out by one's colleagues and given special recognition. By some yardsticks, honors like these make life worth living.

Chapter 5

Beyond Research

Every life is a profession of faith, and exercises an inevitable and silent influence.

Henri Frederic Amiel (philosopher)

From that moment on I knew my profession in life was and has remained until today an actor's life.

Leon Askin (actor)

Civilization is a movement and not a condition, a voyage and not a harbor.

Arnold J. Toynbee (historian)

Stardom isn't a profession; it's an accident.

Lauren Bacall (actress)

The price one pays for pursuing any profession, or calling, is an intimate knowledge of its ugly side.

James A. Baldwin (author)

This profession has fed me creatively and allowed me to have a home life and a private life.

Julia Barr (actress)

I'm learning to accept the lack of privacy as the real downer in my profession.

Halle Berry (actress)

Maybe this world is another planet's Hell.

Aldous Huxley (author)

I was provided with additional input that was radically different from the truth. I assisted in furthering that version.

Oliver North (soldier, political commentator)

Academic battles are so vicious because the stakes are so low.

Henry Kissinger (historian, politician)

5.1   Writing a Textbook

Textbook writing is book writing to be sure, but it is quite a different animal from the other types of writing in the profession. The textbook market—especially at the lower division (i.e., freshman and sophomore) level—is *very* competitive. And everyone teaches lower division courses—this is our bread and butter. So everyone thinks that they have a good textbook in them. As a result, it's pretty tough to get a contract to write a lower division book. Once you get that contract, you will learn that it is quite a regimen to get the book written in publishable form. Gone is the idea that you write your stuff, polish it up, send it in, and then go have a beer. Now you will be wrestling with armies of reviewers and developmental editors and other personnel whose job it is to turn your writing efforts into a marketable product. There are also delicate design issues, questions about the use of color, complicated matters of copyright of figures, legalities concerning material borrowed from other sources, and the list goes on and on.

Writing a textbook can take a big chunk out of your life—both professionally and psychologically. Again, consult the sources [KRA2] and [KRA4] for further details in this matter. It is not something to be entered into lightly. You might be wise to consult a more experienced colleague—one who has been down this path—to get a sense of what sort of life-changing experience you might be entering here.

Of course it *is* possible to make a good deal of money in writing textbooks. Roland Larson (a math professor at Penn State Behrend Campus) has produced textbooks and other educational materials that have grossed three-quarters of a billion dollars. I'll let you figure what percentage of that total was his cut. Larson has his own publishing company now with upwards of fifty employees. Mary Dolciani wrote the textbooks on the New Math in the 1960s that the teachers could actually understand. She had most of the state adoptions for several years. Dolciani has now passed on, but there is a Dolciani Prize and a Dolciani Foundation that uses her bequest for educational purposes. James Stewart has had the bestselling calculus book for a number of years now—and there are 500,000 English-language calculus books sold worldwide each year. So he has made a good deal of money.

Money is one of several temptations to write a textbook. I have one friend who wrote a calculus book to aid in putting his many kids through college. This was an ill-fated mission, and the book did not sell at all. He pocketed his advance, and that is about all the money he ever saw from that multiyear project. My view, as one who has written textbooks, is that you had better have more than pecuniary reasons for investing so much time and effort in textbook writing. Perhaps you want to change the way that a subject is taught. Perhaps you really have something to say. Perhaps you want to have an impact on the profession. I think that those are better motivators, with more lasting value, and will see you through a big project like this with more of your life and your integrity intact.

Upper-division textbook writing is a somewhat more gentle process. You, the author, will have more control of the basic steps, and you will have much more latitude to write what you want. The entire writing regimen will probably take considerably less time than for a lower-division book, and you will certainly go through fewer drafts. The entire time-window for producing the book could be as little as two years, and at the end you will have a nice contribution to the didactic literature to hold in your hand and share with your colleagues and family. Writing a graduate text is an even more gentle activity—it's really writing mathematics, and you can concentrate more on the content and less on the exposition (although you *should* spend time on both). I shall say no more about the delineations among different types of textbook writing here; see [KRA4] for a more detailed discussion.

The bad news is that, in today's market, not so many commercial publishers are interested in upper-division publishing (never mind graduate texts or monographs). A successful lower-division text can sell 50,000 copies per year. A successful upper-division text might sell 5,000 copies per year. It's all in the numbers. The sale of the first several thousand copies of the book will cover the publisher's "plant costs"—these are the salaries of the editors, the typesetting costs, the design costs, the production costs, the warehousing costs, the printing costs, and so forth. And there *will be* more such costs for a lower-division book than for an upper-division book. But after the sales pass that basic accounting threshold, then it's all gravy, and publishers are in business to make a profit. So books that sell more grab all the attention. In the old days (even thirty years ago) a publisher would market a calculus book to make some money and thereby be able to subsidize the more academic advanced book program. Conversely, the advanced books would lend prestige to the overall enterprise. No more. Now every book must stand on its own; it must make a profit. This is a different world, and not a very pretty one for those who want to write advanced, recondite mathematics books.

When I was in graduate school there were more than twenty publishers that would develop and market research monographs. That was because most of the major universities and research institutions had standing orders with all the publishers. Any reasonable monograph had guaranteed sales of 1,000 copies, and publishers simply geared their prices so that after 750 copies they were making a profit. It was a win-win situation for everyone. Today money is tighter, library budgets are cut to the bone, and few institutions have many standing orders. Most libraries pick and choose the books that they want, and there are no longer any guaranteed sales figures. As a result, the number of publishers interested in publishing mathematical monographs has dropped to about five.

Just so with upper-division texts. The number of math majors is flat at best, and not large. The booming majors these days are biomedical engineering, supply line management,[1] psychology, and media studies. That is where the publishers are gravitating. If you go to the annual AMS/MAA meeting (held in January), you will see that the number of mathematical textbook publishers has dropped dramatically (from twenty-five years ago) and that the number that are hawking upper-division texts is few.

The real money is in marketing lower-division texts, and the *big money* is in so-called developmental math. Developmental math is corporate double-speak for precalculus and pre-precalculus. Big state schools have *huge* enrollments in these remedial courses. Teaching gurus like Uri Treisman inveigh against it; Treisman feels that the high schools should teach this elementary material and colleges and universities should teach the real stuff. But the reality of life is that students vote with their feet. They get to college not particularly liking math, but faced with the reality that they have to take it. Their preparation being rather weak, they end up enrolling in remedial courses. So that's where the sales are.

5.2   The Mathematics Curriculum

As students, most of us in the profession went through a fairly standard mathematics curriculum at a good school, and we have taught roughly the same mathematics curriculum at a number of institutions. It (the set of courses for math majors) continues to look pretty much like this:

---

[1] You may have never even heard of this one. It is based on **Amazon**'s business plan of having no warehouses, and drop-shipping all book orders directly from the publisher.

- calculus

- linear algebra

- ordinary differential equations

- abstract algebra

- real analysis

- complex analysis

- geometry

The last two items on the list may be considered optional. I had a very strong undergraduate training in mathematics, and I didn't have either one of those courses (I later learned them on my own).

Many departments now offer different flavors of the math major—statistics, applied, teaching, computational, and the like—and this gives students some flexibility in designing their courses of study. At Washington University the most popular math option by far is the statistics option. Turns out that is because it is a tried-and-true way to avoid taking real analysis.

Even though schools that have kept up with the times are now offering students some variation and latitude in the program, the fact remains that the basic structure of the undergraduate major has not changed in quite some time. And perhaps it should. In part because of the rise of computing, and the blossoming of theoretical computer science as an independent discipline, discrete mathematics has become ever more important. Where does that fit into the curriculum? There are those who will argue that calculus has had its day; it no longer jibes with the problems that are most important in our world. The curriculum should in fact center around discrete math and algebra. Others will argue that the most important thing is logic; after all, the **P/NP** problem is (in the opinion of some influential people) the most important question in mathematics, if not in all of science.

Applied mathematics has a high profile these days, and well it should. It makes good sense that the undergraduate curriculum should embrace applied math—at least to the extent of offering Fourier analysis, partial differential equations, and numerical analysis.

You will note that the outline of the math curriculum provided above does not contain any discrete math or logic or probability or statistics. This is not meant to exhibit any prejudice or disapproval on my part, but it could be a manifestation of my ignorance. The key point that I am trying to make is that there are serious issues to consider here. And somebody should be considering them. The NSF announced about a decade ago that it was funding a new program to encourage math departments to discard their old curricula and develop new ones. I think that the jury is still out regarding the success of that program. But it is bound to produce some interesting and useful results.

You, as a tenured faculty member in your department, may wish to take up this gauntlet. Of course you will want to do this as part of a team, so the first step will be to determine whether anyone else is interested. And you will want to take the temperature of the entire department—perhaps at a faculty meeting—to determine whether your efforts (which will be *considerable* and *protracted*) will be met favorably. *Certainly* find out whether the Chair and the Executive Committee approve of your efforts. You might even ask permission to run it by the Dean.

It would be foolish and disheartening to spend a couple of years redesigning the undergraduate curriculum only to find that everyone is content with the curriculum you have and that nobody wants to hear anything to the contrary. This is the sort of well-meaning but wrong-headed nonsense that can easily lead to major rifts among faculty, and to the sundering of otherwise good and productive relationships. You may think that you are a knight on a white horse, setting out to do a world of good. Others may see you as a misanthrope who just wants to rock the boat.

Of course one path, once you have laid the groundwork for your mission, is to get a grant to support your efforts. As I have already indicated, there is copious government money (and private foundation money as well) to support efforts to beef up the American math curriculum. What you do at your particular school could be perceived as a model for departments across the country. You could find yourself with a $3 million grant—much larger than any other grant in the department—to subsidize your humble efforts. This is, in fact, what happened with some of the teaching reform efforts in the 1980s and 1990s. It was a remarkable social phenomenon—the last shall be first and the first shall be last (I think that the Bible said this first). It had a huge impact on the infrastructure of many departments, and created new resentments that nobody had even suspected. This is an exciting path to pursue, but you should have a clear picture of what you are getting into before you start the journey.

5.3   How to Be a Departmental Citizen

Part of this will be thrust upon you. Most everyone has committee and administrative duties of one sort or another (if you are an endowed Chair Professor then you may be exempt from such activity). These are part of your job, and you should endeavor to acquit yourself professionally and admirably in their dispatch. If you do so, and if it becomes widely known that you are an estimable colleague on committees, one who does his/her homework and contributes constructively, then you will be admired and respected as a departmental citizen.

Of course you can not only embrace this challenge but also carry it further. You can *initiate* programs—be they projects to modify the administrative structure of the department, or to augment or modify the curriculum, or to develop joint curricula with other departments (physics or engineering or finance), or to develop relations (or perhaps an internship program) with local industry. You should certainly clear any such activity with the Chair. (Don't just go off on your own with a bunch of half-baked ideas.) And the Chair may in turn want to run the idea past the Executive Committee. But once you get a green light then you will probably have considerable latitude to develop your own program and put your stamp on something worthwhile. Activities of this kind can count positively toward your tenure or promotion, but don't kid yourself. They don't count as much as writing and publishing research papers. Being the departmental factotum is not the same as being a world-class scholar.

Most of us find that the best way to live our lives is to concentrate on being mathematicians. If we are called upon to serve on a committee, or head a task force, then we do so in a serious fashion and to the best of our abilities. If you are called upon to serve as Chair or in some other administrative capacity, then you should take the matter seriously, and perhaps agree to be a candidate. But most of us don't go out of our way to *assume* administrative tasks. And it all works out in the end. Our primary mission is *not* to serve on committees. In fact our primary mission is to do mathematics and to teach mathematics—not necessarily in that order.

5.4   Letters of Recommendation

Once you have become an established mathematician, you are likely to be asked for letters of recommendation. Such a document could be a letter of recommendation for a tenure case, or for a promotion, or for both. It could be a letter recommending a young person for a first or second job. It could be a letter recommending a senior person for an endowed Chair Professorship, or for the Chairship of a department. (For the sake of this discussion, I will call these "professional letters".) It also could be a letter of recommendation for a student—an undergraduate applying to graduate school, or a graduate student applying for a job, or a student at any level applying for a fellowship (such letters are treated a bit differently from professional letters—see below). There are many variants from which we can distill out some unifying principles on writing letters of recommendation.

In my view, it is both unprofessional and irresponsible to dodge the task of writing a letter. Let me be more precise. There *will* arise circumstances in which you either cannot write or should not write. Perhaps you have had a fight with the candidate in question and feel that you cannot offer an objective opinion; perhaps you have a conflict of interest; perhaps you are unfamiliar with the general area in which the candidate works; or perhaps you do not know the candidate well at all. It is also possible that you know the candidate quite well, but have a poor opinion of him/her.

In any of these cases, or similar ones, you should quickly and plainly write to the person (the Dean or Chair) who requested the letter and say that you cannot write. Sometimes the person who is to be the subject of the letter is standing right before you and asks you directly to write on his/her behalf. Again, you should give this person an honest answer right away. If you can write an accurate and supportive letter in a timely fashion, then say that you will do so *and do it*. If for some reason you cannot write or will not write, then say that. *Do not shilly-shally around*, and do not be dishonest. Don't tell the person that you will write, knowing full well that you will never do it. If you cannot write, best is if you can give the reason, but it is acceptable if you cannot. Do not agonize over the task for six months and *then* decline to write; take care of the matter right away.

There are certain individuals—and I certainly admire these folks for their organization and tenacity—who have set up an interactive Web page for students who request a letter. When a student comes in and says, "Please write a letter for me", these faculty tell the student to go fill out the Web form. It requests the student's name, level, what courses he/she has taken from the professor, what grades were received, what other interactions the student and professor have had, other extracurricular activities that would be pertinent, academic and career goals, distinctions in the department, and anything else that would be relevant and would help this professor to write a good letter for the student. What a great idea! Certainly this reduces the amount of legwork that the professor will have to do; in effect, the student has already generated a draft of the necessary letter. And it provides an ironclad mechanism for keeping track of the letters that you need to write. Many times the professor will set up the system so that it sends him/her timely reminders for writing the letters.

Having decided to perform the task—that is, to write the requested letter—you must do what you have been asked to do. That is, you must formulate an opinion, state it clearly, and defend it. The standard format will be explained below.

In the first few sentences, state plainly the question that you are addressing. For example:

> The purpose of this letter is to support the tenure and promotion of Lloyd George. I have known the candidate and his work for a period of six years, and have been impressed with his originality and his productivity. I indeed think that tenure and promotion are appropriate. My detailed remarks follow.

Alternatively:

> You have asked for my opinion on the tenure, and promotion to Associate Professor, of Dr. Andrew Jackson. Dr. Jackson is now six years from the Ph.D., and in that time has produced nothing but some rotten teaching evaluations and a letter to the editor of the *Two-Year College Math Journal*. Based on that track record, my opinion is that he is worthy neither of tenure nor of promotion.

The bulk, or body, of the letter follows, and it should support in detail the thesis enunciated in the first paragraph. The concluding paragraph of the letter should sum up the case, restate the thesis, and emphasize the conclusion.

To repeat: Once the body of the letter is written—and this could comprise one or two (or even more) pages—then you must write a concluding paragraph. You *must* write it. You must sum up the points you have made, and restate your thesis. A sample of this practice is:

> In view of the stature of Laszlo Toth in the field of computational algebraic geometry, and considering his accomplishments as a teacher and as a scholar, I can recommend him without reservation for promotion and tenure in your department.

(I am assuming that you have in fact described Toth's status and accomplishments, in a favorable manner, in the preceding paragraphs.) Another possibility is:

> In sum, I feel strongly that Ish Kabibble should not be promoted or tenured. Indeed, I cannot imagine the circumstances in which such a move could be considered appropriate.

There are those who, although experienced letter writers, do not adhere to the general scheme just described. One of the standard rationales for this behavior is that, in many states and at many institutions, it is (theoretically) possible for the candidate to have access to the complete text of his/her letters of recommendation—including the identities of the writers. If such is the case, then the soliciting school will inform the writer at the time the letter is requested. Of course the letterwriter is offered the option up front of declining to write if he/she is uncomfortable with this "freedom of information" situation.

There are those who, still uncomfortable, agree to write but are afraid to say anything. The most negative thing that they are willing to do is to "damn with faint praise". Not only does this artifice undercut the responsibility of the letterwriter, but it puts on those evaluating the case the onus of trying to figure out what the writer was trying to (but did not) say. In the best of all possible circumstances, someone at the soliciting institution will phone the letterwriter and just *ask* what the letter was meant to say. In the worst of circumstances, the evaluators are left to guess what was meant. Given that someone's life and career are in the balance, it is a genuine shame for such a circumstance to come to pass.

Enough preaching. I will now give some advice about the body of the letter. If you want your (professional) letter to have some impact, and to be taken seriously, then you must do three things: **(i)** make some specific comments about specific work or specific papers of the candidate, **(ii)** speak of particular qualities of the candidate, specific contributions that he/she has made to the subject or to the profession, detailed information about his/her teaching or departmental service, objective information about his/her academic qualifications, **(iii)** make binary comparisons. You may also wish to discuss other attributes of the candidate. No matter what these may be, you should heed these principles: be *precise*, speak of *particular* attributes, and speak only of those topics of which you have *direct knowledge*. Now let me explain.

Your letter had better say more than "Konrad Adenauer is a hail fellow, well met. Give him whatever he wants." First, such a letter does not say anything. Second, given the circumstances described above, in which some letter writers attempt to avoid litigation by "damning with faint praise", such a vague letter could be construed as sotto voce damnation. If your comments are instead detailed and specific then it is difficult for people to misconstrue them.

Thus you should dwell, for a page or more, on specific virtues of the candidate's scientific work. Make detailed remarks about particular papers: Why is this result important? How does it improve on earlier work? How does the work advance the field? Who else has worked on this problem? This material should not be a self-serving introspection. Remember that most of the readers of the letter will be nonspecialists. Many, including the Dean and members of his committee, will not even be mathematicians. Thus attempt, briefly, to give background and motivation. Drop some names. For example, say that Ignatz of MIT worked on this problem for years and obtained only feeble partial results. The candidate under review murdered the problem. If appropriate, point out that the candidate publishes in the *Annals* and *Inventiones*—and that these are eminent, carefully refereed journals.

If you can, discuss the candidate's other qualifications. Is he/she a particularly good teacher? Good lecturer (you may have heard a conference talk by the candidate)? Inspiring collaborator? Has the candidate been noted for departmental service (you may have discussed the matter with a colleague)? Is the candidate a valued colleague and departmental member? A distinguished member of the mathematical community? Noted for national service? Painting a picture of this person as a working member of the profession can be valuable, and add to the positive tone of your letter.

It is astonishing, but true, that even highly placed people, who write dozens of influential letters every year, seem to be unaware of the need for binary comparisons. To put it bluntly, an important letter that is to have a strong effect *must* compare the candidate being discussed to other people, of a similar age and career level, at other institutions. The comparison should be with people—preferably other academic mathematicians—whose names the informed reader will recognize. Thus, if the candidate is an algebraic geometer and you say in your letter that "This candidate is comparable to David Mumford when Mumford was the same age", then most algebraic geometers will know exactly what you mean and will be extremely impressed; they will in turn explain to their colleagues the significance of your remarks. If, instead, you say "This candidate is comparable to Prince Charles when Charlie was a student at Gordonstoun", then nobody will know what you are talking about—and you can be sure that they will not be impressed.

To come to the point, if you are writing an important letter that you want people to notice, then you must say something like

> The five best people under the age of thirty-five in this area are
> A, B, C, D, E.

In the best of all circumstances, the candidate under consideration in your letter is one of A through E—and you should point out that fact. Alternatively, you could say:

> Two of the best people in this field, at the beginning tenure stage, are Jones and Schmones. Candidate Bones fits comfortably between them. Bones is surely more original than Schmones and more powerful than Jones.

Or you could say that the candidate falls into the next group. Or that the candidate is so good that it would be silly to compare him/her to the usual five best. Say what you think is appropriate. But *say something.* If you do not, then the readers will notice the omission and infer that, between the lines, you are saying that this guy is not any good. Better to say that he is number fifteen than to say nothing at all.[2]

---

[2]A caveat is in order if the letter that you are asked to write is *not* solicited from a research institution. If the candidate is in fact at a four-year college or a comprehensive university or a liberal arts school, where the primary faculty activity is teaching, then the institution probably demands a lot of classroom activity—and not so much scholarship.

Your letter of recommendation can contain other specifics and details that might grab the reader's attention. You could say that the candidate gives excellent talks at conferences. You could say that he/she is a wonderful collaborator. You could say that the candidate has beautiful insights, and that talking mathematics with this person is a pleasure. You could describe in glowing and heartfelt terms the process of proving a theorem, or of writing a paper, with the candidate.

These days, credible evidence that the candidate is a good teacher will certainly help the case. Of course you are probably not in the same department as the candidate, so you very well may not be able to discuss his/her teaching. If the candidate is a truly outstanding teacher, then perhaps you have heard his/her colleagues mention his/her talents, or perhaps you know that he/she has won a teaching award. It makes quite an impression on letter readers if Professor A from University X can comment knowledgeably and in detail on the teaching of Professor B from College Y.

When you are writing a letter for a candidate, then a heavy responsibility rests on your shoulders. The Dean or Chair who solicits the letters of recommendation is not simply casting his/her net and taking a vote: this person wants a *mandate*. He/she will *not* weigh good letters against bad: he/she wants to be socked between the eyes. A tough Dean once told me "If a case is not overwhelming, then I turn it down. If the candidate is any good, he'll land on his feet. If not, then we are better off without him." Thus if your letter for a tenure case says

Benedict Arnold is no good. Don't do it.

then you may as well face the music and realize that *your letter alone* will have killed the case—at least for now. I cannot repeat this point too strongly: It is dead wrong to say to yourself "This is a negative letter that I am writing, but it will not count unless all the other letters are negative too." Baloney! One negative letter will usually stop the case cold. That is all there is to it.

These days, almost every school wants its permanent faculty to have some sort of academic profile; but a teaching college can hardly expect its instructors to stand up to hard-nosed binary comparisons. The lesson is this: Read the soliciting letter carefully; speak to people in *their language*, and tell them what *they* want to know. If the soliciting letter is from a teaching institution, then it is probably most appropriate for you to write about teaching, curriculum, publications in the *Monthly*, and letters to the editor of *UME Trends*. A disquisition on Gelfand-Fuks cohomology is probably less apropos.

The writing of letters of recommendation is not formulaic. Indeed, if all letters of recommendation fit a pattern and sounded the same, or if all *your* letters look the same, then they will eventually be ignored. Mathematicians keep a mental database on letter writers in the same way that good baseball pitchers keep a database on batters. After several years, we know who "tells it like it is" in his letters, who spins tales, and who simply cannot be trusted. We also know who always writes the same letter for everyone. And we act accordingly.

You will develop your own style of writing letters. Mathematics is a sufficiently small world that, after several years, people will recognize your letters of recommendation at a glance. But, no matter how you write your letters, you will want to take into account the issues raised in this section.

Although there is an art to writing a "professional letter", at least you are dealing with familiar territory, and speaking of matters on which you are expert. Any professional mathematician for whom you might write has a publication list, a track record in teaching, a reputation as a lecturer, and some essence as a collaborator. When you are writing for a student, by contrast, matters are more nebulous. The student has none of the professional attributes that you are comfortable discussing. Yet, if you want your letter to be memorable, and to be perceived positively, then you still want to say something noteworthy about the student.

While the precepts of organization that I have stated above still apply in a letter for a student, some of the other particulars do not. For example, you most likely cannot remark on the student's scientific work, and you most likely cannot make binary comparisons. In fact any attempt at binary comparison is likely to be ludicrous.

Thus in practice you must try a bit harder to say something specific about the student for whom you are writing a letter. After you have been teaching for several years, it may be the case that you have actually taught a few thousand students (this would be true, for example, if you have taught calculus in large lectures several times). It becomes difficult to distinguish students—even good ones—in your memory, much less to say something of interest about any of them. If you apply yourself to the task, then you can nevertheless come up with some noticeable things to say. Here are some examples, taken from genuine letters:

> Betsy Ross is one of the five most talented undergraduates that
> I have encountered in twenty years of teaching.

Benjamin Disraeli is hardworking and perseverant. He can think on his feet—at the blackboard—just like an experienced mathematician. He is original and imaginative.

In order to test her creative abilities, I have given Mary Poppins extra work outside of class. She discovered a new proof of Gronwall's inequality, discovered Euler's equation in the calculus of variations on her own, and has also posed numerous interesting problems of her own creation. Needless to say, she breezed through all the standard class work. [I actually once said these very things about a rather remarkable student. He got into Princeton for graduate school.]

As usual, the point is to say *something*—and that something should be quite specific. The view of letter *readers* is that if the letter writers cannot say anything unambiguous and remarkable about a student, then there is probably nothing memorable about that student. So what if the student can earn mostly *A*'s in his/her classes? That is no big deal, and in any event can be gleaned from the transcript.

One of the most critical, and delicate, types of letter that you will have to write is a letter seeking a job for a student completing his M.A. or Ph.D. *under your direction*. Your statements and enthusiasms are a priori suspect because you obviously have a vested interest in finding this student a job, and in seeing him/her succeed. Thus you must strive to put into practice the precepts described above: (i) say why this student is good, (ii) say what this student has accomplished, (iii) if possible and credible, compare the student favorably with other recent degree holders, (iv) say something about the student's ability to teach, (v) say something about the student's ability to interact with other mathematicians.

A meat-and-potatoes job application from a fresh Ph.D. has a detailed letter from the thesis advisor that conforms, at least in spirit, to the suggestions just enumerated. This detailed letter is accompanied by two or three additional letters from other instructors at the same institution, each of which is rather vague and says in effect "Doo dah, doo dah; see the letter by the thesis advisor". If you want your student's dossier to stand out, and to really garner attention, then you should strive to help the student make his/her dossier rise above this rather dreary norm. Endeavor to ensure that the other writers know something about what is in the thesis. If possible, convince someone from another institution to write a letter for the student. Make sure that the dossier includes at least one detailed letter about the student's teaching abilities.

When you write a letter of recommendation, tell the truth. If all your letters read "This candidate is peachy, and a dandy teacher, too. Give him X." (where X is the plum that the candidate is applying for), then after a while (as mentioned earlier) nobody will pay any attention to what you say. I presume that if you take the trouble to write letters, then that is not the result that you wish. The infrastructure has a memory. It will remember whether you are a person who can make tough decisions, or whether you are wishy-washy. If you want your letters to count, then you must call it as you see it. It is hard to be hard, but that is what the situation demands.

One issue that we, as letter writers, often must address is whether or not a job candidate can speak English, and how well (this question could even apply to an undergraduate student—especially if that student is applying to graduate school and might be considered for a teaching assistantship). In this matter we are, in the United States, cursed by our group dishonesty over the past thirty years. Too often have we said in a letter that "this candidate speaks excellent English, can teach well, and is a charming conversationalist to boot". In a more frank mode, we might have said "This candidate speaks better than average English" (recalling Garrison Keillor's statement about the town of Lake Wobegon, in which "all the children are above average"). When the candidate arrived to assume his/her position, the hiring institution often found that he/she could not understand even simple instructions and had no idea how to teach. As with most things in life, it is best to tell the truth about the candidate's ability to speak English, and particularly to communicate (and to understand questions!). You can try to present the facts as gently as possible, but present them you must.

5.5   Writing and Editing

If you do any sort of writing—be it a research paper or an expository article or a textbook or a research monograph—you will end up dealing with one or more editors. If writing becomes an avocation for you, then editors will be a big part of your life. Many of them will become your friends and colleagues. This is a good thing, and will enrich your professional (and personal) existence.

Some editors, such as the "publishing editor" for a commercial publishing house, are typically *not* academics. They are people who were perhaps English majors in college who decided to pursue an interest in scholastics by the means of helping to produce good books. These are people who come from the business side of the picture. Other editors are, in fact, professors who give of their time to help with the publication of a scholarly journal or perhaps a book series. There are yet other types of editors, such as copyeditors, production editors, acquisition editors, and consulting editors. The list could go on. All of these people can impact your life in a variety of ways.

In today's scholarly context, it is possible to think of publishing as simply putting your work (in `*.pdf` or other format) on a Website. After all, that gets it out before the entire world, is fast and simple and free, and is something that you can do yourself. But if you take a broader point of view, if you are worried about archiving, if you are concerned about vetting, if you want to establish your scholarly reputation, then you must think about the publishing process in a broader and perhaps more traditional sense. That is the purpose of this section.

5.5.1 Publishing Editors

A "publishing editor" works for a publishing house or perhaps a professional society. The editor interfaces between the author and the publishing mechanism. That mechanism is either a commercial publishing house or an academic publisher or a professional society or perhaps a freestanding journal.

There is no school or training camp or "minor league" for editors. There is no college major in editing. Most *textbook editors* begin as sales representatives for some publishing house and work their way up through the ranks. Editors for the professional societies, for research journals, and for the high-level monograph series often have advanced training in mathematics or in some science. They do *not* work their way up through the ranks, but are hired straight out as editors.

Your editor is a person with a passion for books or scholarship, who values learning and erudition, and who likes dealing with writers. The editor certainly has an interest in publishing good stuff. But he/she is answerable to the company. The book editor, for example, had better publish more successful books than not or else that editor will find himself/herself out on the street. Likewise, the managing editor of a journal must maintain the quality of the journal, keep the paper flow robust, keep the referees and associate editors in line. Publishing is a business. A lot of what the editor does is businesslike activity: managing people, managing the flow of work, and seeing that projects are completed.

From the point of view of the author, the editor is the person that you go to when you need something or when you have a problem or when you want to make a proposal. The editor is the "point person" for your publishing projects.

A good editor will watch over your work from inception to completion. He/she will help enlist it in the first place, shepherd it through the publishing process, and rejoice with you when it hits the street. He/she will propose and nurture your next project. He/she will be your friend and confidant and advisor.

5.5.2   The Purpose of Editors

Of course the editor represents the publisher to the world. An organization is no better than its people, and publishing is a very human activity.

With today's technology, it would be possible for you to simply shove a TeX disk into one end of a big machine and, after the machine crunched on it for a while, a box of journals or books would come out the other end. Some robot would load it onto a truck or train, and it would eventually be delivered to the distributor. All without human intervention.

The world could some day be run in this fashion, but I am currently grateful that it is not. As has been described elsewhere, people add value to every stage of the publication process:

- The editor helps to attract your book to this particular publishing house. If, instead, you are dealing with a paper or article, the editor may have invited you to submit. Alternatively, you may have selected this editor to receive your work.

- For a book, the editor will design the contract (just for you) and help you to understand it. For a paper, the editor will launch your work into the editorial process.

- The editor chooses the reviewers or referees.

- The editor helps to evaluate the reviews and reports. He/she oversees the rewriting. When there are differences between the author and the referee, the editor helps to clear the air and put matters in perspective.

- The editor decides in which series to publish the book (if it is a book) and in which issue to publish the paper (if it is a paper).

- For a journal, the editor maintains paper flow and quality. He/she oversees diversity in subject matter, overses diversity in authorship, manages page length issues, adjudicates priority disputes, and overall keeps the ship of state on track.

- The editor oversees the copyediting, the art editing, the page composition, and the overall production of the book.

- The editor sees that the book or paper goes to press on schedule, and that it is handled properly after it hits the street.

5.5.3 Types of Editors

No list of the various types of editors can be complete. What we can do here is to describe some of the principal types of editors.

5.5.3.1 Journal Editors

(a) **The Managing Editor** This is the person (usually a fellow mathematician) who oversees the entire operation of the journal—from the academic point of view. The managing editor (ME) deals with the people at the publishing house, supervises all the associate editors (see below), worries about the quality of the papers, worries about the paper flow. The ME makes the final decision about the publication (or not) of each paper. If there are problems with the refereeing, problems with plagiarism, problems with fraud, problems with priority disputes, he/she is the person on the scene.

The ME must keep track of the volume of papers, and must be sure that there are enough papers for the upcoming issues. The ME must maintain quality. And he/she must keep in touch with the people from the publishing house as needed. When necessary, the ME will have to communicate with authors and referees (although this will usually be handled by the associate editors). If a paper is too long in process, if the associate editor is having trouble getting a useful referee's report, or if other difficulties arise, the managing editor will usually have to step in and make things right.

The ME will want to manage paper flow, to be sure that the journal always has an adequate number of papers so that the journal issues can be produced in a timely manner. At the same time, the ME should worry about an acceptance rate. A *low* acceptance rate suggests a journal with high standards; a *high* acceptance rate suggests a journal with modest standards.

(b)   **The associate editors** The associate editor (AE) is answerable to the managing editor (ME). This will also be a fellow mathematician. Typically, the managing editor will assign certain papers to the AE for handling. It is the AE's job to get each paper refereed, to evaluate the referee's report, and to make a recommendation (to publish or not to publish) to the ME (who makes the final decision about publication).

The associate editor must know how to find a good referee for any given paper. There are several parameters to consider: **(i)** the referee should be an expert in the subject area, **(ii)** the referee should be dependable, **(iii)** the referee should not be prejudiced, **(iv)** the referee should be someone who can get the job done. Of course a good and experienced editor knows everyone who is out there and who could be a referee; he/she knows whom he/she can count on and work with.

A good editor must also know how to deal with the referees that he/she has working on papers. He/she must keep track of how long each referee has had his/her assignment (a spreadsheet could be useful here). Managing paper flow, and backlog, is an essential part of running an efficient operation. You may find that using an electronic calendar or personal digital assistant is a good way to monitor work flow.[3]

Different journals have different routines for handling the refereeing process. Some will have the associate editor send a hard copy of the paper to the referee, together with a cover letter (which is usually boilerplate) containing some formal guidelines for the refereeing process.

---

[3]Some commercial publishers, such as Elsevier, have completely electronicized their operations. Today with Elsevier, papers are submitted electronically, assigned to associate editors electronically, and refereed electronically. The editors can use the OnLine system—efficiently, and with considerable ease—to track papers, assign referees, see where papers are in the editorial process, read referee's reports, editorial commentary, and other correspondence, check how long they have been in the system, and to take appropriate action when required.

Others may send an electronic version (probably a *.pdf file) as an *e*-mail attachment, together with an *e*-mail text with referee's guidelines. These guidelines—whether hard copy or electronic—may certainly suggest a time frame for the refereeing process as well. It is not often that the instructions to the referee will give detailed advice on what points to address or how to formulate the report.[4] But many times the letter will say, "Our journal has high standards. We only wish to publish timely papers that are important. Please put this paper in context, tell us whether the results are significant and why." Or words to that effect. The letter to the referee also will usually not suggest a length for the report. But I will suggest one right now.

The report needs to be long enough to do the job. If the paper is short (five pages or fewer), then a referee's report of one page or less will surely be adequate. The referee's report must address a number of key points, and these are outlined in Section 4.10.

Of course a good associate editor must know how to read and assess a referee's report. The ideal report will say, "This paper should be accepted. It meets the journal's standards (which are thus and such) and it has these specific merits." Or the report may instead say, "This paper in no way meets the publishing criteria of this journal and must be rejected. The reasons are ...." Unfortunately, a great many referee's reports are rather more circumspect. Some reports will make some commentary on the paper but come to no conclusion. So the AE must know what to do.

The associate editor or managing editor will want to think carefully about the mechanics of rejecting a paper. The fact that a paper is rejected does not have to mean that the work is bad or the author is a weak mathematician. Papers get rejected for lots of reasons. It is always best if the associate editor can help the author of a rejected paper to maintain his/her dignity and learn from the experience.[5]

These days it is fairly difficult to find good referees who will do the

---

[4]However, some journals will have a form or checklist that asks very specific questions about the paper.

[5]Some journals remove the rejection process from the associate editor's hands. Some central office sends out the rejection letters. I, the author of this book, have no influence over central offices. So I am addressing these remarks to an associate editor (or managing editor) who may have to handle rejections.

job, and do it efficiently and well. My advice would be *not* to burden
your referees with a list of instructions about what you want them to
write in their report, or what form you want that report to take. It is
your job to take the report that is submitted and make good use of it.
If this entails editing or rephrasing the referee's report, then so be it.

(c)   **The In-House Editor** If you are the managing editor or associate
editor for a journal published by a commercial publishing house, then
you will work with an editor who is paid by, and is answerable to, that
publishing house. In the industry, such a person is sometimes known
as a "publishing editor". These monikers can vary from company to
company. The titles "commissioning editor", "acquisitions editor", and
"series editor" are also used. If you are the editor for a journal published
by a professional society, then you will work with an editor who is
employed full-time by that society; and he/she will have one of these
titles, or some variant thereof.

This "in-house" editor will not exercise much say over the academic
aspects of the journal. He/she will have no influence over which papers are accepted or rejected, or how the issues are composed. That
is *your* (the academic editor's) purview. Rather, the in-house editor is
concerned with scheduling, with paper flow, with subscription management, and with keeping track of the money. This editor will usually not
try to encroach on your turf, and you will certainly have little interest in
encroaching on his/hers. Typically, you will have once- or twice-yearly
meetings with the in-house editor to discuss policy issues and management issues. In the interim you will have regular communications by
phone and by *e*-mail.

5.5.3.2   Proceedings Editors

Many of us get our start in editing by becoming involved in producing a
conference proceedings. Your role will be as "editor" of the proceedings.
You will usually work with an in-house editor at the publishing house that
will produce the proceedings.

This is a good mechanism for learning the ropes. For you must determine
how to collect papers, how to get them refereed, and then you must decide
which ones to publish and which ones not. You will interface with other

editors, you will communicate with authors, and you will interact with the in-house editor at the publishing house of your proceedings volume.

This experience will give you a taste of the editing life—both its joys and its sorrows. You will get the satisfaction of putting a slice of the mathematical pie before a focused audience, and of having had an impact on your field that may have lasting value. Your proceedings may be for a research conference, or an education conference, or a consortium about the profession.[6]

A volume of conference proceedings places special demands on the editors. The conference will have been peopled by many of your friends, collaborators, and colleagues. They will all be assuming that their work will appear in this volume. The tradition in mathematics, up until about twenty years ago, was that conference proceedings were not carefully refereed and they were often a repository for technical work that was not necessarily of the highest standard. Times have changed, and those few publishers that will even consider publishing a conference proceedings today will almost certainly demand that you, the editor, have the papers refereed.

You will need to set page limits on the papers for your conference proceedings—and *enforce them.* You do not want a collection of hundred-page-long semi-unreadable diatribes in your volume. You also don't want a bunch of 2-page throw-away papers. Instead, you want substantial papers that stand on their own, and have intrinsic merit. You want a volume that will be a milestone in the subject, and a source for those who want to become acquainted with the latest ideas in this area of mathematics.

The Website

http://www.ams.org/authors/checklist.html

has a useful checklist for the preparation of a conference proceedings. This checklist, in fact, can serve as a useful touchstone for most any book in which the author/editor bears the primary burden for preparation. It addresses such issues as:

- Checking the size of the type block (i.e., the printed matter—without the margins—on the page)

---

[6]There are many different types of infrastructural volumes that you may consider producing. Another possibility is to assemble a volume of contributed papers or essays that are *not* derived from some gathering or conference. This could be *The State of the Art in Invariant Socle Theory* or *Contributions to the Teaching Reform Movement* or *New Ideas on Teaching with* Mathematica. Edited volumes are a world unto themselves, and one that you may enjoy exploring.

- Being sure that the entire manuscript is carefully proofread

- Being sure that all necessary permissions have been obtained

- Being sure that figures have been properly rendered in `PostScript` or `*.pdf` or another suitable graphics format, and at the proper resolution

- Being sure that all grant and contract support has been acknowledged

- Including 2000 `Mathematics Subject Classification Numbers` and `Key Words`.

Certainly the in-house editor can help with some of these more mundane matters. Your job is primarily to handle the scholarly side of things.

The conference proceedings volume should have a strong and compelling introduction. Tell what the conference was about, and what the subject is about. Introduce the papers, and introduce the participants. If the proceedings is dedicated to a particular mathematician, then that will help to give your thoughts a focus. Relate the subject of the conference and the papers to the work of this individual. Tell a little bit about his/her contribution to mathematics, and why it is eminently suitable to dedicate this volume to him/her. Your introduction can be three or four or more pages. It should be fulsome and detailed, but should not become bogged down in extraneous tedium.

Editing a conference proceedings can be a rewarding and worthwhile experience. It is a chunk of work, but the end result can be valuable and much appreciated. In your first go-round, do not try to do the entire job alone. Get a couple of colleagues—perhaps some more seasoned scholars—to help you out. Make it clear that you will shoulder the main burden, but get them on board so that you can get useful advice.

5.5.3.3   Book Editors

Many different types of editors are associated with the process of publishing books. I shall begin by describing the different types of editors who will work for the publishing house. After I describe six types of in-house publishing editors, we shall speak of consulting editors—these are academic mathematicians who work with a publisher to develop (for instance) a book series.

(a) **The Acquisitions Editor** The acquisitions editor is the person who finds authors for the company's book series. Many times such an editor will visit college and university campuses or go to professional meetings. The editor will collar people, talk them up, find out if they have any book projects in the works.

(b) **The Production Editor** The production editor oversees the process of developing a book from manuscript form to the format and medium which is ready to go to press. This process includes copyediting, art editing, page composition, accuracy checking, and many other procedures as well.

(c) **The Copyeditor** Working under the supervision of a Production Editor, the copyeditor reads your manuscript painstakingly and checks it for grammar, syntax, accuracy, and consistency (these days, the copyeditor and the production editor *could* be the same person). He/she will be in constant touch with you to work out rough spots or troublesome passages.

It is a pleasure to work with a good copyeditor. You will derive a lot from the experience, and your book will be much better for the process.

(d) **The Art Editor** The art editor will be an expert on graphic design, on the visual display of quantitative information (see the lovely books [TUF1], [TUF2]), and on the software that is available for producing art for books (this includes Adobe `Illustrator`,® Corel `DRAW`,® Harvard `Graphics`,® `MacDraw`,® `MacPaint`,® `xfig`, and many others). He/she will help turn the artwork that you have supplied yourself (whether these be hand-drawn sketches or semi-professional pictures produced with drafting equipment or graphics that you have created with software, or pictures that you have borrowed from other sources) into publication-quality graphics. The end result will be figures for your book that will reflect well on you and on the publisher.

(e) **The Design Editor** If you are publishing an elementary textbook, then design is a big part of the book's future success. Cover design, page design, page layout, design of each of the page elements (how does a theorem look? how does a remark look? how are figures displayed?, etc.), use of color, use of boxes and other elements for design and display, design of the front and back matter, all play a role in how the book will do in the marketplace.

(f)   The Developmental Editor For an elementary book (certainly not for a monograph) there is usually a developmental editor assigned to oversee the reviewing process, help the author to interpret the reviews, recommend revisions and augmentations to the manuscript, supervise the design, and shepherd the project to completion. The developmental editor will help to oversee the work of the copyeditor, the art editor, and others.

(g)   The Consulting Editor Unlike the first six types of editors that we have discussed, this editor is typically a working academic mathematician. A consulting editor works with the publishing house to help develop a book series. His/her role is **(i)** to attract good authors for the series, **(ii)** to explain to authors how the series works, what the royalty rate is, why the author should want to write for *this* particular series, **(iii)** to act as a go-between for the author and the publisher. Of course it is only the in-house acquisitions editor at the publishing house who can negotiate the contract and discuss particular legalities of the document. But the consulting editor, as a fellow academic, can help an author to become comfortable with the process.

5.5.3.4   Editorial Assistants

Both book editors and journal editors can and will have editorial assistants. I have an editorial assistant to help me with the six journals that I work on. She is a marvel. She keeps track of every paper and every editor and every referee. She knows when a paper has been out too long with the editor or with the referee. She knows when there are problems (with an editor who is not doing his/her job, or an author who is not cooperating, or a referee who is unreachable). She interfaces with the typesetter and the printer and the subscription managing service. She brings any problems to my attention, and *I* handle them. This is appropriate; for it is I who have the authority to whip people into line, and it is I who have the scientific expertise. But she does all the legwork to manage all the information and keep the manuscripts and the schedules at our fingertips. I could not function without her.

Likewise, a book editor uses his/her assistant to keep track of projects and authors and reviews. As with my own editorial assistant whom I described above, the assistant to a book editor does all the legwork and clerical work so that the editor can be on top of all his/her assignments.

It is sometimes true in a publishing house that a good editorial assistant will eventually be promoted to editor. A good editor can be promoted to senior editor or managing editor or executive editor, and after that to publisher.

5.5.3.5   The Editorial Board

Often a book series or a journal will have an editorial board.

For a book series, there will be one or two consulting editors (these are academics who are, together with a designated "publishing editor" at the publishing house, in charge of the series) and then perhaps a dozen members of the editorial board. The members of this board are selected by the consulting editors. They function to attract authors to the series, to select subject areas and topics for books, and to review manuscripts. The editorial board provides the imprimatur for each book in the series.

For a journal, the editorial board is the union of all the editors. *The Journal of Geometric Analysis* has a managing editor (yours truly), a body of executive editors, and a larger corpus of associate editors. The executive editors are figureheads. They suggest policy and serve to attract good papers to the journal. The associate editors carry the burden of the work: They are assigned papers to have refereed. The managing editor oversees the entire operation, and makes the final decision about what gets published and when.

5.6   What If You Are from Another Country?

The United States is currently, and has been for many years, a welcoming place to foreign mathematicians. Certainly we benefited immensely from the influx of European mathematicians prior to World War II. This brain drain helped to put American mathematics on the map. In the late 1980s there was another great influx of Chinese and Russian and other Eastern European mathematicians. Our professional culture and our overall intellectual quality was considerably enriched by this social transference.

If you are a foreign mathematician getting settled at a college or university in the United States, then you will have certain adjustments to make. The nature of our students, the nature (and pace and level) of our classes, and the university structure and administration will all be unfamiliar to you. If you want to fit in, if you want to succeed, if you want to get along with your colleagues, then you need to learn some of the basics of your new environment.[7]

In many European countries, a test is given to all students in high school to determine their future course. As a result, some students are directed to the university and others are directed to trade schools. A consequence of this system is that the universities are less open and more focused than in the United States. European universities often do not have breadth requirements, and students jump right into their majors early on. It is quite common for an Italian math major to take twenty-one year-long math courses during his/her undergraduate math education. I am an experienced, senior mathematician, and I have not had twenty-one year-long courses in my entire life!

The European students I have been describing here will also have had a good deal of math in high school—rather more than a typical American student gets exposed to. So an easy mistake for a foreign-born instructor in our country to make would be to misjudge the level of American students. To walk into a calculus class and assume a degree of sophistication and preparation that simply does not exist. The foreign-born teacher may also assume a degree of maturity, and a level of study skills, that is not present. He/she may give exams that are way too hard. Or show a great impatience with elementary questions. Or appear to be rigid or unbending or unfair.

Of course everyone respects the background and the training and the scholarly standards that your foreign education gave you. But you don't want to be the sore thumb in your department. You want to get along, you want to be someone that the group can depend on, and you don't want to make waves with your teaching. With your research you can, of course, fly as high and as close to the sun as you may wish. Set any standard of excellence that you please. But with teaching you may have to retrain (and restrain)

---

[7]In the year 2007, seven senior mathematicians in Sweden were terminated or relieved of their duties or forced to retire or resign. This occurred at several institutions around the country. Many of these individuals were accomplished mathematicians, and well-known in the international mathematics community. And most of them were foreigners (not Swedes). It appears that the nub of the difficulty in most cases was that these were people who made no effort to become team players. Many of them refused to learn Swedish, were disruptive at faculty meetings, and were constantly lodging complaints with the administration. Swedes are people who value collegiality and cooperation. Even though some of these faculty had been on the scene for twenty years or more, the powers that be decided that enough was enough.

yourself.

There is a well-known story of a German professor who was spending a sabbatical leave at a large university in the midwest. To supplement his income, he taught a course for them. In fact, they gave him remedial trigonometry to teach. He had a standard 350-page textbook written by some well-known American author. He stormed into the undergraduate office at the end of the first week and declared, "OK, I've finished with this book. *Now* what do you want me to do?" Just an illustration of how there can be a disconnect when people come from different backgrounds.

The administrative aspects of life are also going to be different. I have one colleague at Washington University who is from India, and he is constantly baffled by our dealings with the administration. At one point we were trying to compose a planning document for the Dean, laying out our vision of the department's development over the next five years. We on the committee kept saying, "Well, the Dean seems to want this. And if we want to please the Dean, then we should do that. And the administration seems to have those expectations." After a while, in exasperation, my Indian colleague said, "If you already know what the Dean wants, then why do you call it 'planning'?"

Dealing with the college or university administration is something that we all have to get used to. It can be mysterious, it can be Byzantine, it can be frustrating. Talking to each other about it is one way to come to grips with the situation; working with the administration is another. It is an observed fact that very few foreign-born scholars become high-level university administrators in this country. There is just a barrier there that is hard to bridge.

I have a friend who was a notably accomplished mathematician in Texas. At one point in middle age he became a high-level administrator at a university in Germany. This took us all by surprise. We knew that his wife was German, and they had been interested in moving to Germany, so this change made sense in terms of where he wanted his life to go. This man's university had been working with Germany to develop a new German university with a new curriculum based on an American model. And he had been involved in the process. So in some sense the transition was fluid and natural. Nevertheless, it was astonishing to see that he could sell himself to the Germans and succeed in his new position.

The bottom line for the foreign-born professor trying to make his/her w in this country is to be open-minded and to listen. *Talk* to people. Wat television. Learn how things are done. Learn to understand the language at a visceral and colloquial level (this is crucial if you are going to understand your students). Many foreign faculty become *very* popular teachers, and I believe that there are several reasons for this:

- They bring a fresh approach to teaching that the students appreciate.

- They are willing to rethink their attitudes in the classroom and to learn the basic precepts anew.

- They create an atmosphere that the instructor and the students are in this together. They make the learning experience a shared adventure.

- They inject refreshing aspects of their foreign culture into the discourse of the subject and help to bring it alive for the students.

In particular, an alert scholar who is newly arrived to the United States *knows* that he/she is going to expend extra effort to become acclimated to the new environment. Unlike a native American, who could be jaded from the get-go, this person will be striving extra hard to make a good impression. And the hard work pays off.

I will repeat that our welcoming of colleagues from other countries has been, and continues to be, a source of strength for our profession. All of us should pitch in to help them acclimate, and to make them part of who we are and what we do.

5.7   National Service

The AMS (American Mathematical Society), the MAA (Mathematical Association of America), and SIAM (Society for Industrial and Applied Mathematics), to name just three, are large, national organizations. Each has a developed infrastructure for both governance and service. It would be reasonable, and natural, for you to belong to both the AMS and the and perhaps some other mathematical organizations as well). Many cians belong to the AMS and SIAM; many belong to all three or Doing so, you may find yourself asked to serve on some national

ce committees are of many different stripes. The AMS has committees (see http://www.ams.org/secretary), and I em all. Among others, the AMS has:

- an Ethics Committee

- a Publications Committee

- a Committee on Committees

- an Editorial Boards Committee

- a History of Mathematics Editorial Committee

- a Graduate Studies in Mathematics Committee

- an Executive Board

- a Council

- a Board of Trustees

- an Investment Committee

- a Long-Range Planning Committee

- a Salary Committee

- a Nominating Committee

- an Abstracts Editorial Committee

and there are many, many more. It may be difficult, at first blush, to determine which of these committees is interesting or which is important. The Editorial Boards Committee, just as an instance, is important because it determines who serves on the editorial boards for different journals, and who will serve as managing editor. The Nominating Committee is significant because it oversees who is nominated for important national offices.

Bear in mind that the AMS, MAA, SIAM, and all the other active and useful mathematical professional organizations, exist for the benefit of mathematicians. This includes mathematicians of all stripes—researchers, teachers, expositors, and others. The benefits that our professional organizations provide for us are truly valuable, and enrich our lives. Just as an instance, the AMS has been a pioneer and a leader in developing electronic media for disseminating mathematics, and in exploring means to keep the costs of research journals under control. All the professional organizations hold important and useful meetings. Most of them publish mainline journals that are part of the mainstream of the subject. They are part of the fabric that holds our profession together.

Serving on national committees (i.e., committees of some professional organization) is informative and useful, and it can also be fun. It is great to give back to the profession, and also exciting to be on top of things and to know what is happening before anyone else does. Plus you can put this service on your CV, and it will count—at least a little bit—towards your tenure or promotion.

5.8 *A Mathematician's Apology*

Every mathematician's favorite manifest of what it is like (psychologically) to be a mathematician, what we all strive for, what the life is all about, is G. H. Hardy's *A Mathematician's Apology* [HAR]. This book is a remarkably touching and lucid explanation of what Hardy's life was meant to be, and the story travels well. It has charted a course for mathematicians for many years. It is a truly moving journey to read this book, and one that we all should experience.

I recommend this, along with Littlewood's (of course Littlewood was Hardy's friend and collaborator and soulmate) *Mathematician's Miscellany* [LIT], for a good dose of the classical view of our profession.

Chapter 6

Being Department Chair

Civilizations die from suicide, not by murder.

Arnold J. Toynbee (historian)

We mustn't complain too much of being comedians—it's an honorable profession. If only we could be good ones the world might gain at least a sense of style. We have failed—that's all. We are bad comedians, we aren't bad men.

Graham Greene (author)

We have indeed at the moment little cause for pride: as a profession we have made a mess of things.

Friedrich August von Hayek (economist)

It's a business you go into because you're an egocentric. It's a very embarrassing profession.

Katharine Hepburn (actress)

All the world is competent to judge my pictures except those who are of my profession.

William Hogarth (painter)

A friendship founded on business is better than a business founded on friendship.

John D. Rockefeller (businessman, philanthropist)

Remember, information is not knowledge; knowledge is not wisdom; wisdom is not truth; truth is not beauty; beauty is not love; love is not music; music is the best.

Frank Zappa (rock musician)

6.1   What Is a Chair?

Some departments have Chairpersons and some have Heads.[1] A Chairperson is a leader among equals. Apart from a modest salary bump for being Chairperson, the Chairperson is no different from anyone else; he/she is just the designated point person for departmental business. A Head is different. The Head is answerable directly to the Dean and can make decisions math majorously—without consulting the other faculty. The faculty can and will make recommendations to the Head, but the Head makes the decisions.

Most mathematicians think that they elect their Chairperson or Head, and most mathematicians would be wrong. The Dean appoints the Chairperson or Head. Period. The Dean may take the faculty vote as a recommendation, a plebiscite, or even as a mandate. But the decision is the Dean's and nobody else's. The Chairperson or Head serves at the pleasure of the Dean, and can be terminated by the Dean.

In the present book I shall consistently use the word "Chair" to designate the departmental leader (either the Chair or the Head). Any refinement of the term will be indicated in context.

6.2   Characteristics of a Chair

Being Chair of the department is *not* just a social position. It is true that the Chair can give the "welcome back" party in the Fall, can host dinners for colloquium speakers, and can be the standup person at various social events. But there is a lot more to being department Chair than that.

Always remember that the mathematics department *is* the faculty. It is not the building, it is not the library, it is not the computer lab, it is not even the students. It is the faculty. The faculty defines the curriculum, teaches the courses, grants the degrees, does the research, and maintains the flow and direction of studies. The Chair's job is to facilitate all this activity.

A math department has an important pedagogical and scholarly mission. Its charge is to educate the students and also to nurture and promote research, and scholarship, and good teaching among its faculty. The job of the Chair is to make sure that those things really happen as effectively as

---

[1]When Szegő was at Stanford he was the Executive Head of the mathematics department—appointed directly by the University President and the Board of Trustees. Thus he had considerable power and math majory.

possible. This involves much hard work and many skills, plus cooperation from colleagues and staff. It is *not* to be taken lightly.

Being an effective and well-liked Chair demands some people skills. You must learn to be patient. I am proud of the fact that, during my five-year term as Chair, I never yelled at anyone. I never cursed at anyone. I might have yelled and cursed in private, but nobody ever knew about it.

You can disagree with someone without losing your respect for him/her. Any interesting topic can be viewed from many different perspectives. It is possible to draw many different conclusions from the same set of data. You and your colleagues will examine a number of different issues together, will exchange views, and will reach some working conclusions—often by consensus. This can be done honorably and amicably, even though you will sometimes disagree (sometimes strongly). The point is that, at the end, you all must be able to shake hands and move on. The Chair can facilitate this process.

A Chair should be honest and have reasonable, verifiable ethical standards. And he/she should hold others to those standards. Your professional code will reflect both on you and on your department. You don't want to do anything that will compromise either. You may, as Chair, be put in the position of having to call a colleague or a staff member on his/her conduct. You may even have to chew him/her out. Just remember that it is essential to let him/her keep his/her dignity. You are only trying to educate these people in the error of their ways; you are not trying to debase them.

The American Mathematical Society has an Ethics Document which lays out guidelines for our professional behavior. It is a good and well-thought-out document, and one that you as Chair should be acquainted with. It can be found at

http://www.ams.org/secretary/ethics.html

6.3 First, Do No Harm

This medical admonition in fact applies to many aspects of modern life. Your primary goal as Chair should be to do the most good for the most people most of the time. Otherwise, why do the job? There will be times when you are faced with nothing better than a choice among evils, but even then you are trying to minimize the harm caused.

You can take some pleasure and some satisfaction in the job of Chair if you **(a)** strengthen the undergraduate program, **(b)** attract more and better graduate students, **(c)** recruit outstanding faculty, **(d)** manage and develop the curriculum, **(e)** nurture the teaching mission of the department, **(f)** promote research, and **(g)** create a convivial and productive atmosphere in the department. And you can then look back on your period in office as time well spent.

Most of the time, like most of us, you will be concentrating on daily tasks, just trying to get the job done. But there will be occasions when it is appropriate to ponder the long-term goals of your job, what you have been able to accomplish, and what you might hope to achieve in days to come. What do you want to do next for your faculty, and what can you do to facilitate these goals? What do you plan for the math major, and what *specific* policies can you implement to effect those plans? What can you do to strengthen and enhance the graduate program, and who can help you do it? You will have to set aside blocks of time so that you can really give some careful thought to your goals and your wishes for the group. This is one of the most important things that you will do. It is *not* something that the typical working mathematician does on a regular basis; but it *is* something that the typical working Chair should do periodically.

You should always be motivated by the concept of the greatest good for the greatest number. It is almost always a mistake to pander to special interests—even if that special interest is a wonderful mathematician who just proved a great theorem.

6.4   How to Become the Chair

First, it is virtually a prerequisite to being Chair that you be a full Professor. This is in part because you are supposed to be a leader, and therefore you should have some seniority. Perhaps more important is that your job will entail promotions of Associate Professors to full Professor and other personnel questions that virtually necessitate that you have a senior rank in the department.

There are exceptions to what I just said. One of my former Ph.D. students served as Chair of her department when she was still an Assistant Professor. But this was an unusual situation; the department had just two faculty, and they took turns.

In some departments there will actually be a *campaign* for the Chairship. Each candidate will be given various opportunities to make his/her pitch, answer questions, show his/her wares. In other departments the process is treated more dispassionately. The names will just be put on a ballot and a vote taken. In still other departments there will be a single faculty meeting in which all the candidates can let it all hang out, answer questions, strut their stuff. And then there is a vote. I can tell you by contrast that, at some important departments in major universities, there is no vote and precious little consultation of the faculty about who should be Chair. The Dean strikes a deal with one of the faculty to captain the ship of state, there is a handshake, and that's it.

In any event, when it becomes time to select a new Chair, it is common for the departmental members to be consulted in some fashion (in some highly proletarian departments, like U. C. Berkeley, the staff and even the janitors are sometimes consulted). In most cases there will be a formal ballot. Sometimes there will be a departmental meeting at which each of the candidates will make a speech and foment some discussion. Sometimes, in order to sidestep the idea that the department is "electing" the Chair, the Dean will actually sit down and talk to each faculty member in private. This way he/she can assess sentiment and (one hopes) identify a consensus.

When the dust clears, the Dean will issue an announcement saying, "I have asked Professor Delmore Schwartz to serve as Chair for the next five years, and he has agreed." And that is it. There will be some mechanism for the sitting Chair to move out of the royal office, and the new Chair will be moved in, and there you have it.

The term for the Chair will vary considerably from school to school and department to department. A Head (the autocratic type of Head that I described at the beginning of this discussion) is by nature more like a benevolent dictator and is liable to serve for as long as ten years. A Chairperson (the more populistic leader that I described at the beginning of this discussion) is generally understood to be rotating, so the term could be three years or five years. Often the term is renewable. Deans like working with a Head or Chairperson who will be in office for a while. Assuming that this is someone the Dean can get along with, the Dean prefers a seasoned veteran whom he/she has dealt with before, and who is reliable and steady—even, one may daresay, predictable.

6.5   How to Stay Chair

An experienced and somewhat embittered department Chair once said to me that he viewed his life as a finite-dimensional vector space. Each of his colleagues represented a dimension. And every year the dimension of his space shrank. Which is to say that every year there were fewer people in the department who would speak to him. His list of friends was growing ever shorter.

The trouble with being Chair is that your two chief jobs are **(i)** to explain the Dean to the department and **(ii)** to explain the department to the Dean. The Dean, if he/she is an experienced administrator, will probably not get terribly emotional when you discuss teaching loads or salaries or research track records with him/her. But the faculty—especially those with little acumen or experience for administration—are liable to get quite exercised over these same questions. And of course to blame you for any circumstances that do not suit their fancy. You may rapidly become the "bad guy," and be subjected to accusations like, "You are no longer one of us. You are now a stooge for the administration."

This is certainly not what you signed on for when you agreed to be Chair. And surely you never planned to cut down on your Christmas card list. But it is a hard reality of life that people who know full well that they have no control over taxes or social welfare or the war in Iraq do in fact believe that they have, or should have, considerable sway over the direction and future of the math department—and well they should. The progress and development of the math department is, in fact, their bailiwick, and they have a vested interest in seeing the enterprise prosper. If these people feel that you, in collusion with the Dean, are wresting that power from their hands, then they are liable to become exasperated and to behave rather irrationally in the process.

I certainly do not claim to have the magic pill that will help you to address these issues. I can advise you, if you are Chair, to run your department in as open a fashion as possible. Make extra efforts to be sure that everyone genuinely feels he/she has a voice in all important decisions. And even some of the minor ones. Have plenty of committees so that everyone feels they are taking part. Have plenty of Deputies and Vice-Chairs and Associate Chairs so that you can spread the power (and blame) around. Have plenty of faculty meetings.

A really popular thing to do—something that worked well for me when I was Chair—is to have frequent teas.[2] Many of the top departments, such as Princeton and Chicago and Harvard, have tea every afternoon. You can do this too if the resources are available. And of course you, as Chair, should attend the tea. This is everyone's opportunity to interact with you informally, sound you out on key issues, find out what is in the air. It is a great way to keep the information pipelines clear and running.

A good Chair is not one who never makes waves. It is not a person who just rubber-stamps things and spends his/her time trying to make everyone opiated and content. A good Chair has ideas and will start initiatives and make things happen—even if some of the faculty are intransigent and uncooperative. If you are such a Chair, and if you have a good Dean, then the Dean will stand behind you and help you to achieve your goals. He/she will support you if the faculty start to get restless, and even start to complain. But beware: The Dean has a bigger picture in mind; if he/she starts to perceive you as a rabble-rouser, then there will be a price to pay.

It can happen—and this can be quite unpleasant if it happens to you— that there will be a groundswell in the department to depose you as Chair. This means, in effect, that the department will go directly to the Dean and say that you are not a good or effective leader and they want you replaced. They will probably have a laundry list of complaints, some of them valid and some not, and it will be up to the Dean to decide how to handle the situation.

Deans are very sensitive in these matters. The Dean wants someone at the head of the math department, or any department, whom he/she can depend on and who won't create excessive unrest. (The operative word here is "excessive", and the judgment of when the line has been crossed is obviously a subjective one.) The Dean wants to know that he/she can spend a week thinking about other things, being reasonably sure that the math department will chug along without his/her direct supervision. If the members of the department are over in his/her office raising the roof, then this is interpreted as a negative signal.

The Dean could decide to stand behind the Chair, and thereby tell the department to suck it up and get with the program. This is the least likely of all eventualities. The Dean wants to keep the people content. What is

---

[2]The idea of mathematicians having tea together each day was begun by Oswald Veblen at Princeton in the 1920s. He was married to a British woman and thought the ceremony charming. So he instituted it in the math department there, and the custom spread.

more likely is that the Dean will call a faculty meeting at which he/she will preside and attempt to find out what is in the air. Even with or without a faculty meeting, it is likely that the Dean will call for new nominees for Chair, and then the department will go through the exercise of choosing a new major domo. The sitting leader will be retired, and life will go on.[3]

Unfortunately, a situation as I have just described can be humiliating for the Chair who is deposed. I know of situations where the Chair was so discomfited that he retired or quit or moved to another institution. Many of us do not have the flexibility or resources to make such a dramatic move, so instead we must retreat to our offices and spend our time licking our wounds and deciding what to do with the rest of our lives. This can be a career-altering event, though it need not be a career-ending event. Taken as a whole, it's not a very pretty picture.

6.6   How to Cease Being Chair

The simplest and most direct way to cease being Chair is to let your term expire, decline to seek renewal of office, and then be done with it. The department chooses another leader, and on you go. I would say that ninety percent of the time that is what actually happens, nobody's feelings are hurt, and life proceeds.

But, as indicated in the last section, things can go awry. Feelings can be hurt, nerves can get raw, anger can flare up, and then all hell breaks loose. I know of one situation where a young faculty member *proved* that the sitting Chair—a long-term Chair who had enjoyed many successes—had committed fraud by deliberately miscounting the ballots in an election. One can imagine the sort of upheaval this caused in the department. Four years later, the department is still hurting, and there is still a leadership crisis. I know of another situation in which the department Chair was sincere and hardworking, but had an insufficient supply of "people skills". Everything he did seemed to anger his colleagues more and more. That particular department has always been very political, and people served to stoke each others' anger. The Dean ended up pulling this Chair from office before his first year was out. He is now retired, living a private life, and no longer a member of

---

[3]The operative point here is that the Dean doesn't care a whit for the sitting Chair's feelings. Or whether the accusations against him/her are just. The Dean has bigger fish to fry.

the department.[4]

None of this is very uplifting. You don't want it to happen to you, and probably it will not. But being Chair can be a thankless job, and it can be a trial. It can put a strain on your marriage, and it can sap your strength. Certainly it can interfere with your research program. If you are a good departmental citizen, committed to the welfare of your institution and your colleagues, then you may feel that you ought to serve your stint as Chair. But you should go into the job with your eyes open, and exercise some caution as you proceed.

6.7 The Chair's Duties

What does the Chair's job consist of? This will vary from institution to institution. At some institutions—especially the elite private universities in the East—the administration carries much of the load, and the Chair's job is largely ceremonial. At other institutions the Dean gives the Chair a pot of money at the beginning of the year and expects him/her to run the department. Don't worry: If you become Chair, the Dean will make it very clear what your job is and what his/her expectations are. I shall give here a sketch of typical duties. You will probably deal with all of them to a greater or lesser extent.

In this long section I lay out the nuts and bolts of the Chair's job. Who are the people that you manage, and what are the parameters that you manage? Who assists in the process, and who does not? What will be the roadblocks, and what will be the blessings?

It is a long journey, but an interesting one.

---

[4]You yourself should be sensitive to when it is time for you to cease being Chair. If you have stopped being effective, if you cannot garner any cooperation from your colleagues, if your research is suffering, if your personal life is suffering, if you are no longer achieving your goals, if there is a great successor waiting in the wings, then perhaps it is time to move on. Chat regularly with your friends and colleagues about how things look, and what are the prospects for the future. Consult your spouse or partner. Be mature and realistic about the situation. There are other things in life than being Chair of the math department.

6.7.1   The Budget

Your department will have a budget. This will include faculty salaries, staff salaries, money for supplies, money for travel, money for visitors, money for colloquium speakers, money for computers, and some other miscellaneous funds as well. Some of the wealthy elite institutions have special deals set up whereby they leave some faculty slots unfilled, and the Dean gives the department those salaries each year as an unencumbered discretionary fund. This provides the department considerable latitude in planning, in inviting last-minute visitors, in giving a graduate student who finishes late in the year a short-term position, or in holding spur-of-the-moment conferences. You won't see arrangements like this at most public universities.

Money talks and baloney walks. It is a grim fact of life that money shapes much of what we do. And, in an enterprise like a math department, the resources and funds that you have at your disposal will be the unseen guiding hand in every decision that you make. It has been said that Princeton rose to be the preeminent math department in the country because of the considerable resources that were put at its disposal (see [NAS]). The same, no doubt, could be said about the University of Chicago or Harvard. For the same reason, Sacramento City College (a rather good junior college) will never be a powerhouse mathematics institution. Because they have no money. Your job as Chair is to come to terms with the budget that you have and to make the most of it. Some Chairs become particularly skillful at prying extra resources out of the Dean; others become skillful at finding what parts of the budget are fungible, and making the most of the money that is at their disposal. Still other Chairs spend a period of three to five years basking in considerable frustration.

The first thing that you will learn as Chair is that managing a departmental budget is *not* the same as managing a household budget. The sources of money are different, the ways that they can be used are different, the ways that they can be moved around are *very* different, and certainly the means by which they are replenished are different. In your household you can decide not to eat steak this month, and put aside money for a new car. Your funds are as fungible as you want them to be. In an academic mathematics department such a situation does not obtain. You cannot take money from the xerox fund and use it to augment Professor X's salary. You cannot take money from staff salaries and use it to fund a trip for a colleague. There will be a learning process for you to see how to manage this multimillion dollar enterprise.[5]

Some universities have generous alumni and other donors who set up special funds for distinguished speakers, distinguished visitors, faculty travel, and the like. All of these enrich the life of the math department, and certainly make it a pleasure to be a part of such an enterprise. Most departments, unfortunately, live much closer to the bone.

Typically you as Chair will have a not-very-generous fund for supplies (a fund which does not go up much each year), a not-very-generous fund for colloquium speakers, and a not-very-generous fund for faculty travel. You simply have to learn to operate within the constraints that the figures impose. Faculty will come to you all year long asking for money to attend this conference or that function, and you will have to decide how to dole out the money. Probably you will have to put an upper limit of $500 or so on each request, and even then you may be limited in what you can do. Graduate students will also come to you asking aid for travel. Sometimes the Dean of Graduate Studies can help you out with those requests.

Faculty salaries are a more interesting, and perhaps a more delicate, aspect of the budget picture. Everyone thinks he/she is underpaid, everyone is touchy about his/her salary, everyone is jealous of everyone else's rewards. That's just human nature. At some schools the Chair has nothing to say about faculty salaries. The Dean does it all. More specifically, in the Fall each faculty member fills out a form (sometimes called the *Good Works Form* or the *Annual Activity Report*) reporting what he/she has been up to—papers or books written, courses taught, committee service performed, invitations received, awards garnered, grants earned, and so forth. The Dean will evaluate that information and use it as a basis for deciding what raises to give.[6] The Chair may be invited to contribute additional information about each faculty member, or to make pitches for certain individuals. Or he/she may not be.

At some schools, like my own, the Chair is told in the Spring that his/her department will get a 3% bump in the aggregate faculty salaries. Then it is up to the Chair (in consultation with the Dean) to decide how to divvy up this bequest. *Beware:* You do *not* want to give everyone the same dollar

---

[5]I am including here faculty salaries, staff salaries, and operating funds. If you are Chair of a big department, then your aggregate budget could be $5 million or more.

[6]Some Deans will ask the department Chair to use this information to linearly rank the faculty. The Chair may be assisted by a committee in doing so. Nonetheless, this is not a natural or attractive task.

amount for a raise, and you also do *not* want to give everyone the same percentage increase as a raise. If you do so, the Dean will conclude that you have no leadership qualities and that you cannot discern quality from lack thereof. The Dean will attach little significance to your opinions—both in this matter and in other important situations—and your position and credibility will thereby erode.

This is one of the hardest parts of your job. Life would be grand if you could give everyone a good raise every year, but that simply is not going to happen. There are certain individuals who are working hard and being very productive and getting papers in the *Annals* and being invited to speak at the International Congress of Mathematicians, and they must be rewarded. Others have made mighty contributions to the teaching effort or the curriculum or in particular departmental service.[7] Certain other departmental members will be less productive, contribute less to the general weal, and therefore probably deserve less. A cardinal rule of leadership is that you must be able to make hard decisions and you must be able to defend them. If one of your colleagues storms into your office and demands to know why he/she got only a 2% raise while somebody else got 4%, then you must be prepared to explain your actions.

Of course the Dean will discuss each raise with you and may overrule some of them. Or suggest modifications. Or at least suggest that in future years you do things differently. One dodge that you may cautiously employ is to tell your colleague that you wanted to give him/her a higher raise but the Dean wouldn't approve it. You should only use this ploy when it is actually true, and you are still honor-bound to provide the details if you can.

I know one very successful Chair who would sit down privately with each of his colleagues every year and say, "OK, George, let's look at your contributions around here and discuss what sort of raise you should have. I can see that your teaching evaluations are of this quality, your contributions to the curriculum are thus and so, your publication record is at this level, and your overall service is about average. I think that you deserve a 2% raise,

---

[7] Just as an example: Today a big project for a department to engage in is to apply for a VIGRE (Vertical Integration of Research and Education in the Mathematical Sciences) grant from the NSF. Such a grant can provide millions of dollars per year to beef up the graduate program and the postdoc program in a department. Those math departments that have VIGREs have been substantially changed—often for the better. Putting together a VIGRE proposal is a *huge* undertaking, and the faculty member who oversees the process deserves a lot of credit.

and that is what I am going to request for you. If you want more, then we have to think about how you can contribute more. I could, for example, raise your teaching load. Or you could be more productive in your research. Or ..."

You can see that this is really a hard line, and not all of us would be up to implementing such a policy. If you are going to endeavor to follow this path, you should first discuss it with the Dean and be sure that he/she supports you. You cannot very well tell a colleague that, if he/she teaches more, then you will raise his/her salary unless you can actually follow through and do it.[8] The Chair whom I just described is now the Dean at another large university.

Many Chairs—and this is just the *Realpolitik* of faculty life—have a committee of, say, five faculty who adjudicate the faculty raises (this committee might be elected or appointed, and the Dean might be involved in its formation). In the year in which those five serve, they of course have nothing to say about their own raises. This committee evaluates all the *Good Works Forms*, discusses the different individuals and their merits, and makes recommendations. In some departments the recommendation will be streamlined: Each individual will be put into one of three groups—say Meritorious, Good, and Fair. The Chair and the Dean then use these rankings to dole out the raises.

Chairs and Deans like the system of a salary committee because it creates an objective filter for considering faculty emoluments and rewards. If a faculty member comes to the Chair to complain, the response can be, "Well, the committee looked over your credentials and this is what they recommended. You have to remember that there are forty-five hard-working faculty in this department and only so much money to go around. This is the best we can do for you this year." This is not likely to send the complainant away happy, but it is an honest answer to the query.

As you dole out raises, a good rule of thumb to follow is to give everyone at least something. A 0% raise is a real slap in the face.[9] As long as the

---

[8]At some state schools now it is *mandated* by the Legislature that there be a system for regular evaluation of faculty quality and development. The Deans enforce this legislative mandate by telling the department Chairs that they should set up such a system or else there will be no raises. Period.

[9]When I taught at Penn State we had one particularly intransigent faculty member who hated his undergraduate teaching duties. He once loused up one of his classes so badly that the Chair had to pull him from the class (and replace him with me!—and I was supposed to be on leave at the time). Of course, for this particular faculty member,

person teaches his/her classes and does not cause trouble, he/she probably deserves a 1% increase. That may take away a bit from someone else who is eminently deserving, but it is probably best for the overall *Gemütlichkeit* of the department.

The Chair also is, in principle, in charge of staff raises. Typically, at least in a big department, you will have a head secretary, and you might ask him/her to recommend raises for each of the staff members. It is, after all, important for the head staff person to feel empowered, and he/she will probably have more detailed knowledge of each person's contributions than you do. You, the Chair, will handle the raise for the head secretary.

6.7.2   Promotion and Tenure

This is another big, big part of the Chair's duties. The decision whether to give a young mathematician tenure is a life-changing one. Granting tenure sets the person on one career path; denying tenure charts quite a different course. It happens occasionally that an individual who is denied tenure can get the decision reversed on appeal. But that is the exception rather than the rule. Generally speaking, if the decision is negative, then that is the end of the line.

The tenure process has several steps, and they will vary from institution to institution. But they are roughly as follows:

1. The Chair forms a committee to put together the tenure case for an individual. This includes **(i)** soliciting "outside" letters[10] to evaluate the research,[11] **(ii)** collecting information about the individual's teaching,

---

the Chair's action was *not* a punishment—it was a blessing. For now he did not have to teach. His *punishment* was that he got 0% raises for a period of several years.

[10]Different schools will do things differently. The Group I schools will certainly write for outside letters as just indicated, and probably the Group II schools will, too. But at some schools teaching is the main activity, and it is an activity that is internal to the institution. All faculty are expected to maintain some scholarly activity, and to publish, but most of them will not have big research programs or big reputations. It may not make any sense to seek outside letters from famous professors at major institutions. The tenure and promotion procedure must be tailored to the school in question and to its modus operandi.

[11]There is a certain etiquette to soliciting outside letters that, I am sad to say, is more honored in the breach than in the observance. It is good to thank outside letter writers, and express appreciation for their time. It is helpful if your solicitation letter says a little bit about the tenure process at your university, and what are the guidelines. It is also

(iii) reporting on the person's departmental and university service. I shall say more about some of these components below.

2. After having collected all the relevant information, the committee reports first to the Chair and then to the department about the quality of the case. Typically, the committee makes a recommendation for tenure or not. At many institutions, the Chair has the power to veto a tenure case before it is passed on to the next level. This is a very important, but rarely exercised, feature. Many times a department will vote to give an individual tenure because people like him/her, they feel that the person makes worthwhile contributions to the department, and his/her mathematics looks pretty good. But an experienced Chair can look at the dossier and see that it is flawed. Either there are not enough published papers, or they do not appear in sufficiently distinguished journals, or the teaching is uneven, or the outside letters are erratic or inconclusive.[12] The Chair knows that the Dean and his/her committee will take a hard, objective look at this dossier and find many flaws in it. It weakens the department, and it weakens the Chair, to send on a tenure case that is inadequate and that will not fly. This is a juncture when the Chair must consider exercising his/her veto power. It is a tough decision, but one that sometimes must be made. If you as Chair do decide to exercise your veto power, then part of your duty is to explain to your department what you are doing and why.

In any event, once the tenure dossier is complete, the department will have a meeting about the tenure case. Someone (probably a member of the tenure committee for this case) will make a presentation to the faculty about the merits of the case. Any weaknesses will be noted at that time. Often a vigorous discussion ensues. Some department members go out of their way to see the good features in any case;

---

nice to tell the letter writers afterward how the tenure case went (i.e., did the candidate pass?). At some schools, the Dean solicits the outside letters. So the matter is out of your (the Chair's) hands. If the matter is *in your hands*, then strive to be courteous. You are asking quite a lot—in terms of time and effort—from these letter writers, and they will take the task very seriously.

[12]It is essential that each letter say *point blank* that the candidate deserves tenure. Each letter should make binary comparisons with mathematicians in the same field at a comparable level. Each letter should speak in detail about the mathematical work. The book [KRA2] discusses in detail the art of writing a good letter of recommendation. See also Section 5.4.

others will be considerably more critical. Everyone wants to maintain standards in the department, and to tenure only those who will improve the overall quality of life. In the end—one hopes—there will be a consensus either to tenure the candidate or not.[13] Generally speaking, the vote is by secret ballot.

3. If the department approves the tenure case, then it is passed on to the Dean. The Dean will want to know the vote count, and he/she may demand to know exactly how each individual faculty member voted. Typically the Dean will have a Tenure and Promotion Committee, consisting of faculty from across disciplines, that will study each tenure case very carefully, discuss it, and make a judgment. Usually the department Chair will make a presentation to this committee, and will be expected to answer some rather pointed questions. Then the Dean's committee takes a vote. The committee's vote is a recommendation to the Dean; he/she may accept it or not. In the end, the Dean makes a decision about tenure and passes it on to the next level.

4. Most colleges and universities have a Provost who sits above all the Deans. Whereas certainly the department, and probably the Dean, will look at all aspects of a tenure case—especially the research—the Provost takes a broader view. He/she will probably feel that the research has been adequately reviewed by the time a case reaches the Provost's level. So the Provost will want to know whether this candidate is a good teacher, whether he/she is a good departmental citizen, whether he/she will contribute to the overall welfare of the university. If, in fact, this person's tenure is approved, then he/she is liable to be around for forty years and cost the institution $5 million in salary and benefits. Is this a good bet? The Provost, too, will have a committee to help in the decision.

5. Finally the Chancellor and the Board of Trustees must pass on each tenure case. One might think that this is just a rubber stamp. After all, the Trustees are mostly business people and other public figures. What

---

[13]Different departments have different rules in this matter. I can tell you straight out that if the vote is close, then the Dean will not be very happy. He/she wants a *strong* majority in favor of tenure. Some departments require a unanimous vote. At my own university, the ballots are signed and the Chair is required to *explain* each negative vote to the Dean.

do they know about academics? They cannot judge the scholarly work of an Assistant Professor. They also do not know much about teaching. So we might think that they would accept the recommendations of the Dean and the Provost. For the most part they do. But there are exceptions. At my own institution we had a tenure candidate who was a fine biologist, a good teacher, and was well liked by his colleagues and his students. But he was a Communist. Really. He had cut sugar cane with Fidel Castro. The Trustees had a problem with this. The Chancellor and the Board of Trustees had a distinct standoff over this one.[14] In the end the Chancellor prevailed and the candidate got tenure. But this was a situation in which the Trustees had their say.

There are many legalities associated with tenure cases, and the best advice I can give you is to be very careful. The institution will indemnify you, the Chair, so that you are protected if you are sued for turning someone down for tenure. That is, as long as you were not negligent or incompetent. You would be wise to familiarize yourself with the school's tenure procedures and rules. Have a session with the Dean or another appropriate administrator to learn how it is done. Consult a previous Chair to learn the basic steps. Study tenure dossiers from recent cases—both successful ones and unsuccessful ones.

Promotions to full Professor are handled similarly to tenure cases. All of the basic steps outlined above are still relevant. Some of the emphases will be different. A candidate for full Professor will be expected to have a more mature and well-developed research program (the phrase "international reputation" frequently comes up in this context). He/she will be expected to have done more university-wide service. He/she will be expected to have a more extensive and bountiful teaching dossier.

One essential feature of both a tenure dossier and a promotion dossier is the Chair's cover letter. This is a letter of five pages or so presenting the case to the Dean and his/her committee (and to the administrators and the committees above the Dean). This is the definitive document that delineates the case to the world, and it should be prepared with exceptional care. In writing this letter, you must assess all the outside letters, evaluate the teaching, and get a line on the service and other departmental contributions. You must craft an incisive, compelling, cogent presentation of all the candidate's

---

[14]The case of Ward Churchill at the University of Colorado also got the Chancellor and Trustees involved—see also Section 7.3. This was not a tenure case, for he was already tenured. The decision, rather, was whether to terminate him.

qualities. Don't make the letter too long (more than eight pages) or too short (fewer than three pages). You should say just what needs to be said, and not more. Our Dean laments of English and Philosophy Chairs who write eighteen-page cover letters for tenure cases. That is not useful. Nobody has the time or patience to read such a screed.

Be honest in this cover letter. If there are flaws that will be evident to anyone reading the case, then you should endeavor to address them in the letter. If there are particular strengths, then bring them out. The best possible letter will answer any questions that may come up in a reader's mind as he/she evaluates the evidence. The Chair's cover letter can make or break a tenure or promotion case. So it should represent your best effort. After you draft it, show it to your Executive Committee or some of your trusted colleagues. Revise it and hone it and make it the sharp tool that it should be.

Part of the process that I have been describing is that Assistant Professors need regular assessment of their progress *before* they get to the tenure decision. Certainly there should be a three-year review—about halfway to the tenure mark—to tell the junior faculty member how he/she is doing. This should be like a mini-tenure review (some call it a "mid-tenure review"). You should solicit about two outside letters, have a hard look at the teaching, and examine the service component. After all the information is in, after you and your Executive Committee have evaluated it, then *you* must sit down with the Assistant Professor and assess the case honestly with him/her. Engage the candidate in discussion, and make sure that he/she really understands what is being said. It is all too easy for the young Assistant Professor to say, "Uh huh, uh huh" and not really listen to anything that you are telling him/her. This is one time when you had better be sure that he/she gets the message. If necessary, have the person's senior mentor chat with him/her as well.

In *very rare* instances you will say to this junior faculty member, "Bob, this really isn't working out. You should look for another line of work. We shall renew your contract for one more year and then you will be terminated." I know of very few cases in which this has actually transpired.[15] What is more typical is that the Chair will say, "George, your dossier is mostly strong. But there are a couple of items that require attention, and I thought we should

---

[15]If you *do* decide that it is appropriate to terminate an Assistant Professor mid-term, then the Dean will need to be involved. Certainly there will be legalities to consider.

discuss them." And you do. The point of the exercise is to bring the Assistant Professor along, give him/her strength, and make it more likely that he/she will achieve tenure a few years down the line.

Many departments actually have *yearly* reviews of the junior faculty. These are much smaller in scale than the three-year review. There will be no outside letters for the annual review, as one doesn't want to pester and/or waste the time of these famous letter writers. The yearly review would consist of a committee of two senior faculty looking over the CV, the research and teaching achievements, and the service component of this individual's record and then writing a one-page report summarizing their observations. Again, the Chair should sit down with the candidate and go over the good and the bad points. *Make sure* that the candidate understands what you are saying. Chat him/her up every few months and determine whether he/she is implementing what you taught him/her—whether progress is being made.

At my own university, there was a tenure case in the History Department in which the candidate's research credentials depended on her pioneering work in "History on the Web". This was a tough case for all concerned. In the end the department put her up for tenure; they thought she had built a substantial body of scholarly inquiry that constituted a serious contribution to the literature of their subject. The Dean did not agree. Deans, and especially the Dean's committee, tend to be narrowly focused and traditional. They have rigid and hidebound ideas of how one establishes a research reputation, and how one checks and verifies research credentials. To come to the point, they are looking for a body of work published in established journals (most likely hard-copy journals), ones that are carefully peer-reviewed and that have a record of publishing excellent work. They expect the candidate's research agenda to follow a traditional path, using sources that have also appeared in well-established and recognized journals. Doing research on the Web is unfamiliar, its standards are not well-recognized, and there is little track record. *Publishing on the Web* is even more problematic, because nobody is quite sure how papers are reviewed, what the filtering process is, and what the standards of the OnLine journals are.

I have published a good deal, and my reputation is a matter of record. I might choose to publish in an OnLine journal just to show my support for the enterprise, or because I want to get my ideas out there quickly and to a broad audience. But a junior candidate for tenure has to face the matter more analytically. He/she should become familiar with how things are done at the particular university at which he/she is facing tenure. Some schools will be eager to branch out into new lines of inquiry, new forms of publishing, new styles of scholarship. The more established institutions probably will not.

6.7.3   Hiring

As indicated elsewhere, academics take comfort (and empowerment) from the fact that they each play a significant role in the oversight and management and planning for the department. This is a piece of the world over which they have some control. One of the most effective ways in which a faculty member exercises that control is through participating in the hiring process. When the Dean gives the department a position, it is like free money. The department can go out and hire virtually anyone[16] it pleases (assuming, that is, that the position has no strings attached), just so long as the necessary salary and other perks can be assembled and the candidate can be convinced to come. So, unlike many other humdrum departmental activities, hiring is one in which everyone will participate, everyone will have an opinion, everyone will have a pet candidate. It is exciting, invigorating, and even fun. It will impact the future life of the department, as any new person will be someone whom you will interact with on a daily basis.

In some ways hiring is one of the most glamorous parts of your job as Chair. Ten years after you cease being Chair, people will remember your successes (or lack thereof) in hiring. Because new hires have a long-term, daily effect on departmental life. They can increase the profile of the department, can add prestige to the department, can attract other strong people, and can pump up the teaching. They can also be a catalyst for new research, and can have a number of unanticipated effects as well. The only other activity of the Chair that is quite so significant is the handling of tenure cases. And for just the same reasons.

Of course the entire hiring process is overseen by the Chair. Candidates will be brought in to meet the department, give a talk, meet the Dean, get acquainted with the institution, and so forth. Certainly the candidate will have an interview with the Chair, or perhaps with the Chair's hiring committee. Along the way, the Chair will become acquainted with the candidate, and will find out what his/her particular needs are. What salary? What kind of startup fund? Does he/she have a spouse with special needs? Does the candidate want to take the first year off to spend in Princeton or Paris?

---

[16]Of course the Dean signs off on all new hires—*especially* tenured hires.

The Chair will ask some pointed questions, and endeavor to determine just what this particular candidate has to offer. What can he/she contribute to the research program, to the teaching effort, to overall departmental development?

After the Chair has assessed the candidate and his/her needs, then it is necessary to consider the case that must made to the administration. Can the university and the department muster the wherewithal to meet all the candidate's demands? Will the Dean participate constructively, and help you make the offer? Will he/she meet the salary needs? The startup needs?[17] Or will he/she be an albatross around your neck, making it a foregone conclusion that you lose the person you really want and instead will hire someone well below the standards that you have set for the department?

Usually one point of contention is salary. Deans like to advertise jobs at the beginning Assistant Professor level. This is partly because such a job represents a minimal commitment (*no tenure*) and also a minimal salary. Often your candidate will be at a more advanced level. If the candidate is actually a somewhat more advanced Assistant Professor, then the Dean will probably not give you a very hard time about it. But if the candidate is a tenured full Professor, then you will have a fight on your hands. This is not at all what the Dean had in mind. You must make a case that this is the chance-of-a-lifetime for the department. That the entire nature of the enterprise will change—and definitely for the better—if you can hire this very special person. If you can make a case that Princeton and Harvard also want to make this scholar an offer, it will strengthen your case. Of course such a candidate is liable to want a princely salary, and that will give you problems with the Dean as well. You must argue that, if your institution is going to be competitive in the marketplace, then it must pay competitive salaries and offer competitive perks. You as Chair will have to learn a whole new vocabulary for selling candidates to the Dean.

You may want to give some thought to what hoops you want a visiting job candidate to jump through. Of course he/she will meet the hiring committee, and probably the rest of the faculty as well. If the job is a tenured position, then the candidate will certainly meet the Dean. It is standard for the candidate to give a ceremonial talk and then be taken to dinner. At my own

---

[17]Startup funds are easy from a fiscal point of view because they are one-off items—i.e., nonrecurring debts. The salary is, of course, an entitlement, and will certainly give the Dean pause for thought.

institution—at least for a time—we required every job candidate to give a calculus lecture *in front of a group of calculus students*. This may have been a good idea in theory, but it really irritated the heck out of our job candidates, and it did not seem to be an effective use of anyone's time. It might be more propitious to have a job candidate give a "student colloquium".

I have, as a job candidate, been asked to meet with a teaching committee and to discuss current teaching issues (calculus reform, self-discovery learning, computer-aided learning, and so forth). I have been asked to go drink beer with the other analysts (this was not too taxing). I have been asked to go on a hike in the woods. I happily complied.

It is just a fact that emotions can run high at hiring time. People feel that the future of the department, the quality of life, and the ranking of the program, depend critically on each hiring decision. Especially these days, when money is tight and there is not nearly as much hiring as there was forty years ago, each new position is perceived as something precious and irreplaceable. Each faculty member will not only defend his/her own candidate vigorously, but may also attack your candidate. In many instances you will find yourself comparing apples with oranges,[18] and it will be difficult to see objectively which candidate is truly "best" for the department or the university.

Some departments will have a Long-Range Planning Committee that, among other things, considers the hiring plan for the department. The advantage of such a group is that it can take a broad view and tend to smooth out the approach to hiring questions.

The message here is that hiring is one context in which your people-management skills will play a central role. You will have to supervise some big egos and some hurt feelings and some lousy behavior. You will have faculty stand up at meetings and say in effect, "I am clearly much better than you and I should be listened to and you should not." It is your job then to point out the merits of all sides of the argument, to make sure that everyone has a voice, and to guarantee that all candidates get a fair hearing.

Looking back on my own time as Chair, I feel that some of my biggest and most lasting accomplishments are my successful hires—new people who came and stayed with us and have had a powerful and positive effect on departmental life. If you do things right, then you will be able to take some

---

[18]Frequently a department will have vigorous discussions about which areas to emphasize in a search, or whether the department should seek the absolute best candidate regardless of field.

of the same satisfaction in these activities.

One of the time-consuming, bureaucratic aspects of dealing with a hiring situation is making peace with the university's Affirmative Action Committee. The purpose of this government-mandated committee is to see that hiring is evenhanded and above board, to ensure that members of underrepresented societal groups get a fair shot at any job that is offered.[19] This means that you must prepare an Affirmative Action dossier for each job that you advertise. You must provide statistics that show the nature of your search, and what steps you have taken to make your process fair and inclusive. This can create a great deal of extra work for your staff (special spreadsheets, extra data accumulated on each candidate, special ads in particular periodicals). But you must take this exercise seriously. I know of job searches on which the Dean has pulled the plug—and this was very late in the game, when the department had a candidate to which it *really* wanted to make the offer—because it was perceived that Affirmative Action was not handled correctly.

It has been true for a long time that a new faculty member in physics, chemistry, biology, or another lab science will have a startup fund. This is because such a scholar will likely have a lab, and it takes a good deal of money to set up a new lab from scratch. The candidate will no doubt get grants to keep the lab going from year to year, but the initial configuration of the lab—that is, the *startup*—will require a preliminary infusion of funds. This could be quite a lot—half a million dollars or more. If there is specialized equipment, it could be a couple of million dollars. And Deans are accustomed to this. Especially if the Dean is a chemist, he/she will know what a mass spectrometer or a chromatography system costs and will be prepared with the old checkbook.

The idea of a startup fund for a mathematician is relatively new. Most of us don't have labs. What we do have are computing needs, needs to travel to meet with a collaborator, needs to bring in visitors, needs connected with getting one's Ph.D. students set up. So it is common now for the Math Chair to ask the Dean for a startup fund for a job candidate. The sad fact of the matter is that the Dean is less sympathetic to this request than one

---

[19]The intention here is to fight the "old boy network". As an instance of that network in action, I got my first job—a very good Assistant Professorship—with one phone call from my advisor. No letters, no teaching evaluations, no nothing. Just a phone call. Today this would be considered wrong—and it should be. The most important feature of hiring is that all candidates get a fair hearing, and have a fair chance at the position.

might hope. First of all, Deans like to save money. Second, the Dean just doesn't understand our needs as well as he/she would understand the needs of a lab scientist. Many times the Chair can pry $20,000 out of the Dean for a startup fund for an "ordinary" faculty member, and perhaps $50,000 for a "superstar" faculty member. Other times not, and you as Chair will have to scrape something out of the (already limited) departmental budget.

It is also common to offer a reduced teaching load (at least for the first year or two) to a new faculty member. This is just one more perk that you can throw the candidate's way to make the job offer more attractive. If you are lucky, the Dean will support you in this effort. More likely, however, you will have to do this "under the table" by shuffling everyone's teaching assignment around a bit.

Finally—and this is a big issue these days—there is the so-called two-body problem. Many a job candidate will have a spouse or significant other who also has a career and therefore has professional needs. Sometimes that significant other will be a mathematician, and you will find yourself looking at two hires instead of one. (Many Deans are anxious to pay lip service to their commitment to solving two-body problems. But when push comes to shove—when the Dean may have to produce another position out of thin air—then you may find that he/she sings a different tune.) Section 3.10 discusses two-body problems.

6.7.4   General Departmental Management

On a day-to-day basis, the Chair must face a variety of tasks. These include oversight of the budget, managing the staff (and refereeing internecine squabbles), requisitioning supplies, writing reports for the Dean, engaging in a variety of planning activities, dealing with faculty needs and problems, supervising the teaching and research, and many other jobs that come up as needed.

Many Chairs find it useful to have an Executive Committee (and in fact many Departmental Bylaws mandate that the Chair have one). This is a group of about five senior faculty with whom the Chair meets regularly to hash out decisions and set the course for the department.

The Executive Committee can be a useful tool for a canny Chair:

- First, it is always useful to have trusted colleagues off of whom to bounce your ideas, from whom to get advice, and from whom to gather other perspectives.

- Second, there will be times when you just don't know what to do. A particular situation has you completely flummoxed. A good Executive Committee can help you wade through the complexities and to see what is really going on.

- Third, the Executive Committee can serve as a buffer. If someone—a colleague or an administrator or a student—comes to you complaining of some decision, you can say, "The Executive Committee considered the matter and here is what they decided and why." That takes some of the onus off of you, and spreads the responsibility around in a plausible manner.

- Fourth, when used with restraint, the members of the Executive Committee are people whom you can ask to help out with key tasks. You can't ask them to do large tasks on the spur of the moment. But if you need a little legwork done, or some vital information, then they often can help.

Typically the Executive Committee is elected by the department. But some slots on the Committee may be appointed—by the Dean, or by the Chair himself/herself. Usually a new Executive Committee is appointed each academic year. You will generally find the Executive Committee to be a useful and constructive ally which will make your job a bit easier.

In a reasonably good-sized department there will be a secretary dedicated to undergraduate affairs (and who works closely with the Vice-Chair of Undergraduate Studies) and another secretary dedicated to graduate student affairs (and who will work closely with the Vice-Chair of Graduate Studies). These people are your trusted agents, and frankly will know more about the details of their respective purviews than do the cognate Vice-Chairs. You should consult with these folks regularly just to keep track of the programs and to be sure that the curriculum is functioning as it should.

Also check regularly with your computer people to see that the system is running properly, that there are no crises in the offing, and that the equipment needs are being met. Consult regularly with the faculty to see if the system is being used effectively and well. What new needs are in the offing? Of course your Computer Committee can be a great help in these matters.

No reasonable Dean is going to leave a department without *e*-mail or without adequate computer support. If you plan ahead, then you can often get the Dean to help you when there is a "sudden" need for a RAID system or a laser printer or a multibus server.

Finally, keep in touch with your departmental Librarian. As discussed in Subsection 6.7.25, the library is the intellectual heart of the department. Today, with electronic media, the library is becoming more and more distributed. But this probably means that the managing issues (and the costs!) are more complex. So you want to be well-informed in the matter.

Finally, stay in touch with your faculty. Ralph Boas, when he was Chair at Northwestern, used to enter his office by climbing the fire escape and then squirming through the window—this to avoid his colleagues. My view is that this is not a healthy attitude. Especially if you are a Chairperson (rather than a Head), it is important that you maintain contact with your colleagues and have your finger on the proverbial pulse of the department.

6.7.5   The Reward System

Several years ago the AMS formed a task force—supported financially by a big grant—to study the reward system in the profession. The question was: Are mathematicians fairly rewarded for their efforts? Do all the big raises and all the perks and all the promotions go to those who prove the big theorems? What about efforts for the curriculum? What about departmental and university service? What about extraordinary teaching performance? How about service to the profession?

The AMS task force visited a great many departments and interviewed department Chairs, Deans, and rank-and-file mathematicians to get a sense of what was in the air. Certainly the general feeling among the rank-and-file was that the reward system was not equitable. You can imagine that, in today's American math department, most of the denizens are in their forties or older. Likely as not, their best theorem-proving days are behind them. But they still want to be rewarded. They still have families, kids to put through college, and other tangible needs. So they want raises and other creature comforts. They feel that they are still making valuable contributions to the departmental weal, but not getting a fair amount of recognition. Chairs and Deans acknowledged that there was a problem here, but were cautious in saying too much about it. The task force recommended that departments rethink the system, but not much came of it.

The fact is that recognition of the American mathematical research enterprise is rather recent. It dates back to G. D. Birkhoff and Norbert Wiener, so is not even one hundred years old. And we are proud of it; not likely to turn our collective backs on it any time soon. It is also a fact that the way you get tenure, and the way you get promoted to full Professor, is to publish. Even at a school where the primary mission is to teach, the powers that be expect each candidate to have a scholarly profile—which means that he/she has to create some new mathematics.[20] So research has a prominent status in American mathematics. Research gets rewarded.

I would say that in recent years the teaching reform movement (see [KRA1] and the references therein for some details on this phenomenon) and the copious federal funding that it has received have somewhat changed our view of matters. People now recognize that the Harvard Reform Project, the Duke Reform Project, and many others are worthwhile enterprises that have had a positive effect. And, at least in principle, they acknowledge that these should be suitably rewarded.

But I can tell you that even very recently, in my own department, a seasoned veteran who is quite sympathetic to the importance of teaching and departmental service was quoted as saying, "I just got a paper in the *Annals* and you are going to give a big raise to these guys who do a good job teaching calculus? Where's my piece of the pie?"

The fact is that there is only so much money to go around, and if you give more to one person, then you are going to give less to someone else. You, as Chair, would be making a strong statement—and the *wrong statement*—not to recognize a paper in one of the top journals. But if we are going to be objective, then we ought to recognize that most mathematicians spend ten to fifteen years doing frontline research (if they are lucky) and much of the rest of their careers doing other things. So we should be prepared to recognize, and to reward, those other things.

One phenomenon that a modern Chair must wrestle with is what we call "salary compression". Namely, in order to remain competitive, departments must offer higher and higher salaries to beginning faculty (i.e., Assistant Professors). These are often *much* better than the starting salaries that many of the senior faculty will remember from when they were hired.[21]

---

[20]The papers [SHA] and [WEN] discuss the balance at a primarily teaching institution of research versus teaching.

[21]As of this writing, I know young mathematicians—fresh off a two- or three-year postdoc—getting offered salaries in the low $70K range to begin an Assistant Professor-

And new faculty these days also get startup funds, reduced teaching loads, and other perks. Of course the time of recruitment is when a faculty member has the greatest leverage, and he/she may as well exercise it. But the upshot of all this is that many senior faculty—especially those who are entrenched and perhaps not too much involved with the further development of the department—will become rather resentful. As Chair, you don't want any of your faculty to be resentful. Because you want to get meaningful contributions—on a regular basis—from all of them. And, of course, those senior faculty will have a vote on the tenure cases of those junior faculty who came in on the big salaries. You can see that some brinksmanship will be needed to chart these waters.

The Carnegie Foundation has played a very positive role in helping people to understand the many different contributions that college and university faculty make, and the different ways that these activities can be assessed and rewarded. In particular, the book [GTM] (sponsored by the Carnegie Foundation) lays out their key ideas.

You will have to work with your Dean to find a fair way to carry out the programs that I have briefly described here. After all, it is he/she who allocates the resources and approves the raises.

6.7.6   When One of Your People Wins a Prize

If one of your faculty wins a prize, then it's time to celebrate. Certainly knock on the individual's door and offer hearty congratulations. And don't leave it at that. Ask about the prize, for what work it was given, and how it came about. Really make a do over it.

It should be stressed that prizes are not just for research and big theorems. Our list of math prizes in Section 4.23 shows that there are major prizes for teaching, for service, for expository writing, and for textbooks. All these prizes are great honors, and should be recognized and celebrated accordingly.

Organize a departmental tea so that the entire mathematical family can celebrate the event. Make a little speech. Perhaps get some senior faculty member in the honoree's field of expertise to make some remarks as well.

You should certainly make an annotation in the prize winner's personnel file. See to it that he/she gets a raise in recognition of this notable achieve-

---

ship. And I also know very senior full Professors (not necessarily at the same institution) with an annual salary of $75K. This situation is certainly not equitable, and it is getting worse.

ment. If you don't make the right sort of fuss over a prize winner, then you will be adding to this (obviously very talented) person's incentive to pack up and leave. I know far too many Chairs who (out of either envy or just plain laziness) do little in recognition of prizes, and they end up paying for it.

Jack Welch was the very successful CEO of General Electric for twenty years, and is considered to have been one of the most imaginative and innovative managers in history. He advises finding any excuse to celebrate with your people. Make them feel important and special. Hold a ceremony and make a fuss.

The advice that I am giving here is perhaps even more important if one of your staff wins a prize. Most universities have a variety of awards for staff. These are nothing like the Fields Medal or the Bôcher Prize, but staff get little enough recognition and appreciation. So this approbation will be important for the recipient. Now is the time to really put on the dog: Show this person how much he/she is appreciated as part of the mathematics department family.

Of course you as Chair may as well face facts: Your prize winner probably wants more than just a tea. He/she is probably expecting a substantial bump in salary, and perhaps other perks as well. The sort of person who wins one prize may win several, and you will have to come to grips with the attendant demands. Such is life.

6.7.7 Relations with Other Departments

Some departments—Earth & Planetary Science (EPS) is a good example—interact naturally with many other units on campus. People in EPS naturally have to talk to physicists and astronomers and chemists and perhaps even engineers. Mathematicians—pure mathematicians—do not. Traditionally, a pure mathematician does *not* collaborate with people in other departments. In fact it is only a recent development—in the past thirty years or so—that a pure mathematician will even collaborate with people in his/her own department.

That is our culture, and there is nothing wrong with it. But we pay a price. Deans value interdisciplinary programs and interdisciplinary curricula. This is in part because such activities advertise well: When Mr. and Mrs. Jones come to campus with their beloved offspring to determine whether *this* is where Mary should spend her formative years, they are impressed by cross-disciplinary science and technology programs, by joint medicine and engineering majors, and by biotech labs and supercomputer centers. It is also the case that interdisciplinary programs tend to attract funding—if only because there are more sources that they can turn to. Such enterprises tend to leave most mathematicians in the dust and, frankly, don't interest us much in the first place.

But, to repeat a shopworn phrase, we suffer for this isolation. If the Dean perceives that we are not team players, that we are not getting with the current flow of instructional development, then he/she is not going to be sympathetic when the Chair comes to him/her for more funds, or more positions, or more graduate assistantships, or more resources of some other kind.

In my travels around the country, I have noted that today mathematics departments are making a concerted effort to correct the image that I have just described. Everyone is aware that, about twelve years ago, the Dean at the University of Rochester shut down the graduate program in their mathematics department. This was in large part because of his perception that the math department was isolated, was not making a viable contribution to the teaching effort, and overall wasn't carrying its weight. With the aid of the American Mathematical Society, the situation at Rochester was finally rectified. But a dire knell was sent out to the mathematics community: *This could happen to you.* In my own negotiations with our Dean at Washington University (when I was Chair), the Dean brought up the Rochester situation from time to time. And he did *not* do so in an effort to be friendly. He was threatening me and my department.

As indicated elsewhere in this book, it's generally a bad idea for the math department to be an isolated entity on campus. We do a tremendous amount of lower division teaching and other campus-wide service. So it is natural for us to be involved in campus life. And we should be.

When I was department Chair, we had a consortium of Science Chairs that met every month or two to discuss matters of mutual interest. Some of these discussions involved asking the Dean for certain tangible considerations:

- Setting up an office to manage Undergraduate Research projects (these are often funded by grants, and there is a considerable amount of infrastructure and paperwork).

- More support of our graduate programs.

- Setting up a high-speed Internet system for the Sciences.

- Providing supercomputing resources.

- Making more of the income from the endowment available to the departments.

Obviously there is strength in numbers. If just one department Chair goes to the Dean and asks for a high-speed Internet hookup, then it is just too easy for the Dean to brush him/her off. But if all the Science Chairs go to him/her as a group, then that is bound to garner some attention. It would be natural to ask whether our consortium ever did anything independent of the Dean. The answer is "no"; our main effort was to present a united front to our administration.

In addition, it is fun and collegial to interact with leaders from other departments. This is your opportunity to learn how other departments deal with some of the same issues—personnel problems, promotions, money for travel, computing resources—that you must deal with. You will find that even nearby fields like physics have quite different values, and quite different goals, from the ones that we are familiar with. And you will learn from discussing them.

I was once in a conversation with other department Chairs and the Chair of Psychology (Psychology was included with the Sciences because it was a very analytical department) said that he thought that the best place for a budding Psychology Professor to go to college was a small liberal arts college. I was a bit surprised. I think that small liberal arts colleges are great— certainly one of the unique strengths of the American education system— but they have definite limitations in mathematics education. A really bright student who races through the undergraduate curriculum in two years will want to start going to seminars and taking graduate courses. A liberal arts college may only have limited choices to offer in this regard. If the liberal arts college happens to be in Cambridge, Massachusetts, then the student can go down the block to Harvard or MIT for some enrichment. If, instead, the liberal arts college is in Portland, Oregon then the choices may be fewer.

It turns out that the reason that the Psychology Chair made his assertion is that (in his view) a student at a small liberal arts college will have lots of opportunities to get involved in a Professor's lab and thereby have hands-on research experience. Well, this had never occurred to me. It doesn't necessarily make sense for mathematics—although there are many liberal arts colleges these days with quite strong math faculty who can offer students a lot. Certainly this is an instance in which interacting with faculty from other departments really taught me something.

6.7.8 Vice-Chairs

If you have a department of reasonable size, then it makes sense to have certain people appointed as Vice-Chairs in charge of particular activities.[22] Many departments have a Vice-Chair of Undergraduate Studies and a Vice-Chair of Graduate Studies. There can also be an Administrative Vice-Chair. Each of these people is a Professor, and usually each of these people is hand-picked and appointed by the Chair. These folks can lift a considerable administrative burden off the Chair's shoulders, and can help ensure that critical parts of the departmental mission are executed smoothly. The Administrative Vice-Chair can carry a lot of the weight in preparing tenure cases, hiring cases, and so forth.

It makes sense to give your Vice-Chairs certain perks as a reward for their services. This could include a salary boost, or time off from teaching, or perhaps a new computer or a travel fund. What is feasible, and what you can actually manage, will depend on what resources your department has at its disposal. In some (unfortunately rare) instances the Dean will actually help you in this matter. In most cases, you are going to have to scrimp and do some creative bookkeeping.

Your Vice-Chairs are also people whom you can consult—much like the Executive Committee—about departmental issues, about matters of policy, about hiring and promotion, and so forth. These should be among your trusted advisors, people you can rely on and who will support you when times get tough. Your relationship with the Vice-Chairs is a symbiotic one, and you can derive much from having these allies.

6.7.9 Committees

If you are a reasonably well-organized department Chair with an agenda, and seem to know what you are doing, then there is a temptation to do

---

[22]It must be noted, however, that many smaller colleges have a math faculty of three or five or eight. If you are Chair of such a department, then it probably doesn't make any sense to have a bunch of Vice-Chairs. On the one hand, you will conduct the majority of your business by consensus. On the other hand, you will actually carry out most tasks single-handedly.

everything yourself. After all, you know much better than anyone else in the department what is going on on campus, what the Dean is thinking, what resources are available, and what the institution's goals are. Why waste your time explaining any of this to your colleagues? They will only muddy things up with their uninformed and puerile questions.

This is an easy attitude to fall into, but it is not the right one. Academics have a very powerful sense of self-determination. They *want* to think that they control the future of the department. On any issue that has any weight, they want to be heard.

And that is the key point. In the end, the buck stops with you (the Chair). But everyone wants to have a voice in the department's future. That is why you need to have committees and regular faculty meetings. Any decision that will affect the whole department should be made by a committee. After all, a tenure or promotion decision affects the whole department indirectly. And, anyway, it is important. It is good to have more heads in on the job.

It is natural to want to put your buddies on all the committees, but you really need to spread the responsibility around. And be sure that junior faculty also have adequate representation on the committees. Junior faculty can easily convince themselves that they are disenfranchised, and will thereby become disaffected. That is not good for anyone. Try to be inclusive in everything that you do.

6.7.10 Graduate Students

As you ponder the great panorama of your life as Chair, you will see certain milestones. Among these may be hiring, tenure cases, promotions, and faculty raises. Perhaps the graduate students loom somewhat smaller in the great scheme of things.

But, in fact, the graduate program is an essential part of any research mathematics department. Graduate education is important, and it requires a good many resources. It is significant that the university funds it, and it does so for a reason.[23]

Training the next generation of scholars is a critical part of what we do. In addition, working with graduate students is *really* good for your scholarly

---

[23]Running a graduate program is *very* expensive. Many countries simply cannot afford to offer graduate education. Prior to twenty years ago, there were no Ph.D. programs in Italy, for example. The great diversity of graduate programs in the United States is one of the things that makes us special.

work. It helps to keep you alive and vigorous.

So it is important to be sure that the graduate students are well treated, and that they know they are well treated. Do your utmost to be sure that the graduate student stipends are at a competitive level.[24] Work with the Vice-Chair for Graduate Studies to be sure that everything possible is done to attract the very best graduate students.

Some departments will actually fly in a group of prospective graduate students each Spring and do a hard sell to convince them to accept the offer and enroll in the program. This may include a dog-and-pony show of various faculty presentations, social time spent with contented graduate students who are already in the program, and time spent getting acquainted with the institution and the city. As Chair, you can make sure that things like this happen, and you can participate to good effect.

The graduate students should know that they can bring you their concerns and that they will be met with an open mind and a can-do attitude. You *want* the graduate students to feel that they can depend on you.

Most of us are aware that, at some schools, the graduate students (at least the teaching assistants) are unionized. This phenomenon seems to occur primarily in the public universities. But when the graduate students at New York University (a private school) unionized, the other private schools took notice. My institution—Washington University in St. Louis (which is intensely private)—took great pains to get together with graduate student leaders, take the temperature of the situation, and see what their concerns were. I myself was quite pleased to learn that the graduate students reported that they were very happy, they felt they were treated well, and especially important to them was that *people in power listened to their concerns* (which ranged from having a voice in what they taught to health insurance coverage to effectiveness of thesis advisors at finding jobs for their students) and acted accordingly. I was proud of my institution. You, of course, want to be proud

---

[24]For many years Princeton University has had one of the top mathematics graduate programs in the world. Many of the leaders in the mathematics community were educated at Princeton. Princeton has never lacked for graduate school applicants, and students literally line up for the privilege of studying at Princeton. But, one year in the late 1970s, Princeton admitted its usual cadre of fifteen or so new graduate students—and nobody showed up! Certainly the Princeton math department was shocked and dismayed, but also confused. It turned out that nobody had been paying much attention to the stipends that the department offered, and they had fallen way short of the norm. Of course steps were quickly taken, and the matter remedied.

of yours, too. So look after your graduate students.

6.7.11 The Graduate Vice-Chair

As indicated elsewhere, you as Chair will ask one of your colleagues—probably a senior faculty member—to direct the Graduate Program. This is an important position, and entails a fair amount of (episodic) work. The Graduate Vice-Chair oversees the admissions process for new graduate students. He/she also supervises the graduate students who are actually in the program. This means that he/she makes sure that the students get through their qualifying exams in good order, get hooked up with a thesis advisor, and make good progress towards the degree. He/she makes sure that the graduate students dispatch their duties (teaching and so forth) in a professional and competent manner. And the Graduate Vice-Chair oversees the payment of graduate student stipends.

If your department puts on a yearly extravaganza for students who have been admitted to the graduate program—in an effort to get them to accept your offer—then the Graduate Vice-Chair will probably organize that effort. Of course his/her secretary will help.

One of the big jobs of the Graduate Vice-Chair is to oversee the admission process for new graduate students. In any given academic year, applications will stream in during the Fall, and perhaps even into January. Many of these will be hard copy, but today many applications come in over the Internet. The Graduate Vice-Chair must work with the Graduate Committee to study and rank the applications. Careful attention must be paid to **(i)** the quality of the school from which the applicant is applying, **(ii)** the quality of the undergraduate curriculum that the applicant has studied, **(iii)** the applicant's performance in course work, **(iv)** the applicant's performance on the Graduate Record Exam (GRE), **(v)** the letters of recommendation.

Another key part of this process is that the Graduate Vice-Chair must see to it that suitable financial aid is obtained for the graduate students. If a graduate program is to be competitive, then it must offer a competitive level of financial support. This is particularly true if the program wants to attract domestic students.

The Graduate Vice-Chair must also pay attention to the balance, among the new admittees, of foreign versus domestic students. Often the foreign students will look more attractive because they will have more advanced training. Some will have already published. But American institutions have a commitment to educating American students (this is especially true if the department has NSF or Department of Education money for its graduate program). Also graduate students usually have to teach (in their roles as teaching assistants). And American students, being native English speakers, have a clear edge in this process. At my own home institution, the Dean of the Graduate Program has mandated that a certain percentage of each new graduate class be native-born Americans—and for this very reason.

Monitoring the progress of graduate students is another big part of the job. It is all too easy for a graduate student to flounder—to get lost in the system, or be overwhelmed by the work. Even talented graduate students run into personal problems, or get depressed, or lose heart. You must be prepared to counsel these people and get them help as appropriate. Also it will sometimes happen that a graduate student will screw up—not meet his/her teaching duties, not do his/her course work, or fall down in some other aspect of assigned or expected duties. It is the Vice-Chair's job to call the person on the carpet and get things straightened out.

This is a lot of stuff. It's a big job. Of course the Graduate Vice-Chair will be assisted by a Graduate Secretary and a Graduate Committee. But the ball is carried by the Vice-Chair. These days the Vice-Chair also has to think about good strategies for recruiting top students to the program. This can include advertising and also asking faculty who are on travel to put in plugs for the program. The Vice-Chair should do his/her best to see that the graduate program offers competitive stipends. These days there is stiff competition for the best students—especially American students. Some departments, even the top ones, offer *signing bonuses* to new graduate students (on top of a generous stipend and a lowered teaching load). The new GAANN grants (Section 4.11) have set a new standard for graduate student stipends, at least for American citizens.

6.7.12   Undergraduate Math Majors

You should look after your undergraduate majors for many of the same reasons that you should look after your graduate students. But there is an even more fundamental reason to do so. Deans pay careful attention to which departments and programs have the majors and which ones don't. If your department has a significant number of majors (say in the top ten or so departments), then you will have the Dean's undivided attention.

When I was at Penn State, we developed some notably flashy ways to attract majors. In those days (the early 1980s), there were more computer science majors than you could shake a stick at. The Computer Science Department kept raising the bar, and it was turning away hundreds of students every year. So the math department developed a computational math major. Well, the number of our majors then skyrocketed!

In a different direction, we noticed that there were a lot of students floating around who couldn't decide what to major in. So we designed the "roll your own" math major, which allowed the student (with proper guidance) to design a custom math major to suit his/her goals and needs. This was another big hit.

With these two devices we increased the number of math majors by a factor of ten. And I can tell you that this was well worth doing. Our profile with the Dean was transformed overnight.

You may find the two examples above—of ploys to attract majors—to be a bit too Madison-Avenue-like. Well, fine. Design your own. There are many ways to attract majors. If word gets out that the math department provides particularly good counseling or advising, then the number of majors will go up. If the department offers a menu of different styles of math major that the students find appealing, then the number of majors will go up. If the department has a vigorous and far-reaching minor program, then the number of clients will go up. Have faculty make a pitch to their vector-calculus classes, or their ODE classes, or their linear algebra classes about the value of being a math major. Invite good students to math department events. Send out e-mails to prospective majors.

It is up to you, the Chair, to tend your own vineyard and make things happen. Of course your Vice-Chair for Undergraduate Studies will help, but it is you who calls the shots.

6.7.13 The Undergraduate Vice-Chair

As indicated elsewhere, you as Chair will ask one of your colleagues—probably a senior faculty member—to direct the undergraduate program. This is an important position, and entails a fair amount of constant work.

Of course the Undergraduate Vice-Chair looks after the math majors, makes sure their needs are being met, and endeavors to recruit more of them. He/she looks after the majors at graduation time, helps to choose some of them for awards, and probably attends the ceremony.

The Undergraduate Vice-Chair also worries about the undergraduate curriculum (see Section 5.2). Is it meeting the needs of the students? Is it contemporary and competitive? How does it stack up against physics and engineering? Does the math department's course requirement and prerequisites make it a tough department, an easy department, or somewhere in between?

In Subsection 6.7.18, I provide a discussion of doling out teaching assignments. This is a very important aspect of departmental life, and one in which the Undergraduate Vice-Chair will take an interest. You may assign him/her to actually formulate the assignments (although you should play a distinct role in the process). After all, it is this Vice-Chair more than anyone else who understands the needs of the majors and the structure of the curriculum.

The Undergraduate Vice-Chair may have a number of miscellaneous tasks, ranging from seeing that the Putnam Team is coached, to ensuring that the honors students get suitable honors advisors, to looking after the undergraduate lounge. In a reasonable world, he/she should be able to delegate many of these tasks.

The Undergraduate Vice-Chair will have an advisory committee that helps with many of these tasks. This committee will meet regularly and consider issues and problems and policy matters attendant to the undergraduate curriculum. But there is a good deal of day-to-day work that is the lot of the Undergraduate Vice-Chair. He/she needs to see that there is a Web page for each course, that syllabi are posted on the Web, that all faculty have office hours, and that those hours are posted.

The Undergraduate Vice-Chair may also oversee undergraduate advising—though the department may have a special officer for that task. In any event, it's part of the curriculum, so he/she will certainly be involved.

It is likely, though not always so, that the Undergraduate Vice-Chair will field student complaints about math courses or math instructors. This is not always a pleasant task, especially if the department has people who tend to generate complaints. It can be time-consuming and aggravating to have to deal with the matter. In some departments the Chair does it; in others it's the Undergraduate Vice-Chair. The Undergraduate Vice-Chair is also likely the point person for cases of cheating. You may already know that, if you catch a student cheating, then you are ill advised to act alone in the matter. You should gather all the evidence and present it to the Undergraduate Vice-Chair. In turn, he/she will probably then present the case to the "Dean of Cheating" (not the official name of this administrator, but the name in practice). The main point here is that the alleged miscreant has a great many rights that must be rigorously respected. You should use the Undergraduate Vice-Chair as a foil in these matters.

As my Dean was constantly reminding me when I was Chair, undergraduate teaching is the math department's most important product. You want to choose a good and hard-working person to be the Undergraduate Vice-Chair. And you, the Chair, will meet regularly with this person so that you can keep an eye on the mathematics program at your institution. You will actually take part in many of the decisions, and you will help to implement them. Part of your legacy as Chair will be the effect that you have on the math major.

From the point of view of the students, and to some degree the Dean, the Undergraduate Vice-Chair is the "point person" for the department. He/she is highly visible, and represents your team's most important product. So choose somebody good for the job.

6.7.14 The Curriculum

The curriculum is the aggregate of all the mathematics courses—and allied courses in physics, computer science, engineering, and the like—that are part of the mathematics education at your institution. Quite frankly, at most schools the curriculum does not get nearly the attention that it deserves. Most of us take for granted that we live with a collection of math courses, most of these are quite familiar because they are pretty much the same courses that we took when we were students, and we take turns teaching them. It is not often that a new course is created, and it is even less common for an old and moribund course to be taken off the books. See Section 5.2 for further discussion of curricular issues.

We probably should, as a regular exercise, review the math curriculum and rethink it. After all, the world changes. We are now teaching more and more non-majors as a service to other departments. The advent of computer science as a freestanding academic discipline must cause one to rethink the role of discrete mathematics and algebra. Cryptography is now an important part of pure mathematics. Statistics is now a much bigger and more robust subject than ever it was before. Most every math department now has a group of applied mathematicians, and applied mathematics is a part of every undergraduate curriculum.

It probably doesn't make any sense to throw the entire curriculum out the window and create a new one. Parts of it—real analysis and algebra and geometry—are just as valuable today as they were one hundred years ago. But the way that the pieces fit together, and the overall shape of the subject, deserve attention and rethinking. You as Chair may consider asking the Vice-Chair for Undergraduate Studies to form a task force to study the matter and make recommendations.

Comments like these could also be applied to the graduate curriculum. In the old days we thought of our graduate students as clones of ourselves, and we trained each of them to be a chip off the old block. But, in today's world, the mathematically trained face a bewildering array of career choices. Government research centers, private industry, defense think tanks, schools of many stripes, the financial sector, the genome project, the Social Security Administration, and the National Aeronautics and Space Administration (NASA) all employ Ph.D. mathematicians. It's fine to teach our graduate students measure theory and functional analysis and differential geometry and ring theory, but perhaps we should offer them more choices. Perhaps we should make it possible for them to work computer science, or optimization, or systems science, or mathematical physics into their program.

This circle of ideas does not necessarily entail that we math faculty must retrain ourselves. It is probably not realistic to expect someone who has been studying Moufang loops for the past twenty-five years to suddenly retool him/herself as an expert in Karmarkar's algorithm. But what we do need to do is to work cooperatively with faculty in other departments who *do* have the expertise in these other areas. We must pay attention to new and rapidly developing fields in the mathematical sciences. Certainly a student can have more than one thesis advisor, and the advisors do not all need to come from the math department. It can enrich us all to have a productive working relationship with engineering or physics or many another department, and it could lead to new research projects as well.

6.7.15   Getting Along with the Staff

The staff—the secretaries, and the computer support team, and the people who tend to the undergraduate curriculum—are in many ways the people who hold the department together. They don't win a lot of prizes nor garner a lot of prestige for the university, but they get the work done. While the faculty are running off to Paris or Taiwan or Rio de Janeiro, giving fancy talks and drinking sangria, the staff are at their desks making sure that the department runs smoothly and reliably. If you are Chair, then your life will be made infinitely simpler if you have a reliable and congenial staff that you can work with and depend on. Conversely, if you cannot get along with your staff, then your life will be hell.

Make an extra effort to be nice to your staff. Of course be friendly and tell them jokes and laugh at theirs. But take them to lunch, bring them little gifts. One particularly popular and successful Chair that I know used to quite regularly bring in a half-gallon of ice cream and ice cream cones and then sit around having a treat with his staff. They loved him for it.

Some faculty are downright rude to the staff. Presumably they think they are so important, or so harried, or so overworked, that they have the right to disrespect those who are less privileged than themselves. This is intolerable, and as Chair you *should not tolerate it.* If a faculty member gets out of line with the staff, then take him/her to task for it. One of the most important attributes of a good leader is to stand up for his/her people. And they will respect the Chair for it and appreciate his/her support.

Treat your staff with respect. Ask their advice, and *really listen to it.* Let them do their jobs without micromanaging; show your appreciation—in some detail—for what they do. Make it clear that you *know* what they do, you *understand* what they do, and you *like* what they do. Everyone wants to be valued. They will give as good as they get. And it is your job as Chair to make sure that the staff is supported and appreciated.

One of your jobs as Chair is to hire new staff. This may be the one task in your panorama of duties that is furthest from your recognized expertise. Generally speaking, the university will screen applicants for university jobs, put their dossiers into some kind of standard form, administer typing and other clerical tests, and then pass relevant folders on to you. My university also hires an independent company to do various background checking on candidates. But, in the end, you will have to interview the candidates and make a choice. It will probably be helpful to have another faculty member at the interview, and probably the departmental Administrative Assistant or Office Manager (i.e., the head staff member).

You will have to think about what questions to ask at the interview, and this will depend in turn on what you are looking for. What technical proficiencies are needed? What people skills? Is the written, official job description an accurate picture of the job? Or are other duties expected? Will this person have to be flexible about hours, or perhaps work on weekends (if the candidate is a single parent then this could be very relevant). Find out what the candidate wants to ask *you*; this could be quite revealing. Of course acquaint yourself with the candidate's employment history, and try to get a sense of why he/she left the previous positions.

In the end you will have to make a decision, help train and acclimate the new employee, and try to help him/her fit in and do the job. Another unfortunate feature of being Chair is that you must also decide when a staff member has to be terminated.[25] A bad apple among the staff can be extremely disruptive and affect the entire departmental work flow. It is your job, and basically yours alone (in concert with the Administrative Assistant or Office Manager), to monitor the staff situation and make sure that things are going smoothly and people are getting along. If they are not, then you must do something about it. There are people in the administration whose job it is to help you with these matters; you should call on them without hesitation to help you get through these situations smoothly.

6.7.16   Working Together as a Team

Getting mathematicians to work together on a project is like herding cats. I am not the first to utter this witticism, and I won't be the last. We are by nature—in fact, by our very training—bohemians. We set our own goals and pursue our own truths. A good geophysicist once told me that in *his* discipline they do *not* value ideas and originality—in fact they find such artifacts quaint and confusing. What they value is organizing data. Such a credo is bewildering and almost offensive to the mathematician. Ideas and originality are just about the only things that we do value. And we tend to pursue them in the tradition of the single-combat warrior—alone on a steed looking for a grail.

However, the view of university administrators is rather different. Deans like departments that work together on big projects. I can't tell you how many times, as Chair, I went to our Dean with some big idea for our department—perhaps reinventing our calculus sequence, or redesigning the math major, or setting up an internship program for our graduate students. Every time—*every time*—the Dean would bestow upon me a pitying smile and say, "How are you going to pull this off with that group of clowns you

---

[25]As Chair, I attended a meeting of Chairs, hosted by the Dean, which he began by saying, "You can't get rid of your tenured faculty but you can get rid of your staff. Take a hard look at your staff. If any of them is not performing up to snuff, we can show you how to terminate them."

have over in the Math Building?" The Dean, too, knew that we were bohemians each charting his/her own path, but he saw the matter somewhat less charitably than you and I do.

I once served on the Provost's Advisory Committee at my university. Our job was to examine difficult situations that the Provost was called upon to adjudicate, and to offer our advice. Examples were **(i)** the possible closing of the dental school, **(ii)** a particularly egregious and thorny case of cheating by an undergraduate, and **(iii)** a tricky tenure case that was being challenged. The tenure case involved a young fellow who had been denied tenure in some engineering department. He thought the decision unjust, and he filed a formal grievance. On our investigation, it did appear that the candidate had reasonable scientific credentials, and his teaching was at an acceptable if not stellar level; so perhaps there was something to investigate. Here is what we found.

This was an academic department that functioned according to the following paradigm (quite unfamiliar to the academic mathematician): The department, as a group, would identify a substantial, long-term research project to work on. It would apply for several large grants to fund the work. Getting the grants, the department would then set up laboratories, apportion the work among the entire faculty (and the graduate students), and then begin their investigations. The point that I want to stress is that *the entire department*—faculty and graduate students and staff—would engage in this big research project. There were no exceptions, no crusty individualists who said, "I've got this other little thing I want to do. The rest of you have fun." It was a zero-one game.

Well, this tenure candidate came from a cultural background wherein it was customary to be rude, abrasive, and uncooperative (by Western standards). He couldn't get along with anyone in the department. He couldn't function usefully at the weekly team meetings. He couldn't collaborate. He was *not*, by any stretch of the imagination, a team player. And *that* is why they denied him tenure. The decision was *not* a judgment against the quality of his scientific work, nor a criticism of his teaching, nor a criticism of his departmental service. They just didn't like him. He couldn't get with the program.

I, as a mathematician, had a problem with this situation. I am quite accustomed to people who cannot get with the program, who are not team players, who revel in being a thorn in our collective sides. Some of these people are my best friends. If I want to form a judgment about a mathematician, I look at his/her record on `MathSciNet`, assess the published works, read some of them as appropriate, and reach some conclusion *based on the quality of the scientific work.* You would never catch me saying, "This person has an abrasive personality, he/she is no fun on committees, he/she can't get with the program in the lab, so don't tenure him/her. Or don't promote him/her. Or don't give him/her the perk that he/she is applying for." I'm just not made that way, and I think that in fact most of us mathematicians are not.

In the end the Provost's Advisory Committee decided to uphold the department's decision not to tenure this particular scholar. Even though the case was problematic, even though there were valid arguments on both sides of the question, in the end we realized that we could not veto the way that this department does business. Clearly this candidate knew what he was getting into when he accepted this job. And he just couldn't play the game. It was unavoidable to conclude that the situation we had was at least partly of his own making. So we upheld the denial of tenure. And he is now gone.

Part of my message here is that the Provost was quite sympathetic to the point of view of that particular engineering department. The Provost values teamwork, and admires a department that can organize itself to work on a big project. He had little sympathy for a candidate who could not toe the line. Like it or not, that is how administrators see the world.

If you are lucky, in your own department, you as Chair will be able to identify a notable subset of your faculty who *can* function as team players. It would be especially effective if some of those team members were also people who were active researchers and good teachers. Because that will give the team credibility, and increase the likelihood that you can recruit new people to the team in the future. Having such a team, you will be able to take on some larger projects, and thereby curry the favor of your Dean or Provost. And, of course, do some good for the department as well.

6.7.17 Supporting Your Faculty's Teaching Efforts

Clearly this is one of the most important things you can do as Chair. When you are just an ordinary citizen in the department, there is little of substance that you can attempt in order to show support for your colleagues' teaching efforts. But as Chair you can make a difference.

For one thing you can reward those who excel at teaching. When it comes time for raises, single out those who shine. Praise them to the Dean—with concrete information to support your case—and propose a good raise. Deans typically don't know much about mathematics, but they fancy that they know a good bit about teaching. They should respond constructively to this sort of proposed emolument.

What else can you do for your teaching force? Provide the equipment that they need. Overhead projectors are easy (because they are relatively cheap). Computers and allied peripherals somewhat less so just because they tend to be expensive. But the department ought to have two or three nice notebook computers that people can use in their classrooms. Of course you need computer projectors to go with them, and the good news is that these have come down dramatically in price in the past couple of years.

Every college and university these days has a program to make classrooms electronic. This can mean providing built-in computers (with a high-speed Internet connection, of course) with projectors and screens and other nice peripherals. It can mean SmartBoards® or variants thereof.[26] Do what you can to get some of the classrooms in the math building—especially the rooms over which you have scheduling control—equipped in this fashion. If your building has an undergraduate or graduate lounge, see that these are outfitted with suitable equipment.

If a faculty member comes to you with a request for support of his/her teaching efforts, be as helpful and constructive as you can. Perhaps he/she needs extra grading time, or another teaching assistant, or wants to make a field trip. Obviously you are not Donald Trump, but you do have some discretionary funds.[27] Try to use them as creatively as possible.

If one of your faculty wants to start a new course—in discrete dynamical systems, or circle-packing, or cryptography, or anything else of hot and current interest—then give suitable support to the project. You probably don't have the expertise to give the project your wholehearted a priori support, but you can see to it that the idea gets a fair shot at being established.

Likewise, if one of your colleagues has a well-thought-out idea for changing the curriculum—say merging linear algebra and ODEs, or creating a new

---

[26]Here a SmartBoard is an electronic display board that you write on with a special stylus. It automatically records what you write as a binary file. Conversely, you can prepare a lecture at home and then present it on the SmartBoard.

[27]When I was Chair, one of my faculty developed a quite innovative course on math and music. This was just great, and I was all for it. But he needed a synthesizer and other high-tech equipment to make the course go. I really broke my back to find the funds for this activity.

discrete math track—then be sure that the idea gets a fair hearing. Talk to the Vice-Chair for Undergraduate Studies to make sure that the right people get to review the proposal and make a measured judgment. The same remark applies to new graduate courses, though there should be some discretion to allow for different people's research interests.

6.7.18   Teaching Assignments

Certainly one of the important things that a Chair does is to put together the teaching assignments. This can be thought of as part of the reward system: The worthy faculty are given the plum courses to teach, and the less worthy faculty are given the more humdrum assignments. A faculty member who does no research and who is only a mediocre presence in service activities might be assigned to teach calculus day in and day out. A faculty member who publishes in top journals, and who has lots of fancy invitations and grants, may be assigned advanced seminar courses and special courses for the math majors.

One feature of a math department, as opposed to many other departments on campus, is that we tend to be rather open-minded about what we are willing to teach. An analyst will teach a broad variety of analysis courses (real analysis, complex analysis, functional analysis, perhaps even numerical analysis) and perhaps some other things, too (partial differential equations, applied mathematics, even geometry). This is part of what makes the mathematical life rich, and in any event is a necessity because of the great variety of courses that a math department offers.

Open up your university catalogue and examine the chemistry curriculum. There is organic chemistry, physical chemistry, radiochemistry, and not much more.[28] Mathematics has many more dimensions, much more depth and texture. The math department has only so many people in it, and they have to cover the curriculum. So we all must be flexible in what we can teach.

Especially if you are Chair of a large department, you may delegate the task of parceling out the teaching assignments to one of your colleagues. This could be one of the Vice-Chairs, or some other specific individual. Of course

---

[28]To be fair, chemistry has expanded its purview in recent years. Especially in view of the interface with biological sciences, you will now find polymer chemistry, bioorganic chemistry, organometallic chemistry, inorganic chemistry, materials chemistry, and the chemistry of biological membranes. Still, it is clear that organic, physical, and radio- form the nexus of the curriculum.

you should oversee the process to make sure that it is done correctly and that the right people are rewarded for their efforts. It is common to distribute a questionnaire to the faculty *before* the assignments are made, asking people what their teaching preferences are and for any other particular requests that they may have.[29]

The manner in which you set the teaching assignments sets a particular tone for the department, and gives people a sense of where your priorities are. Teaching is a big part of everyone's day, so you are impacting people's lives with the way that you handle this job. Give the matter extra care and attention, and it will pay off for everyone.

6.7.19 Supporting Your Faculty's Research Efforts

There is hardly anything more important that you can do in your role as Chair than to identify and recognize and nurture your faculty's research efforts. When you are an ordinary citizen, you can support your colleagues' research by going to seminars, by collaborating with them, by helping them with their grant proposals, and by always being willing to talk. But as Chair you can do more.

Certainly your Dean ought to provide you with a travel fund for your faculty. Be as generous as you can in subventing trips. If a faculty member tells you that he/she needs to go to Paris and you offer $150, that is really not very helpful. Try to suit the largesse to the situation.

Sometimes one of your faculty may want to bring in a visitor—perhaps an important collaborator. Maybe he/she wants the visitor for a full semester. This is going to be expensive. My department has a "reserve fund" which consists of a gratuitous rebate from the administration of grant overhead. I could sometimes dip into this fund for larger requests like this. But many department Chairs will not have such discretion.

If a faculty member wants a new printer, I think that the answer should always be "yes". Printers are cheap, and they make people happy. For a new computer, people may have to wait their turn. If somebody wants something

---

[29]These special requests may take many different forms. A faculty member may ask not to teach precalculus—and he/she may give a cogent reason for this request. He/she may ask to do all his/her teaching in the Fall so that the Spring semester can be free. Or the person may have medical reasons (certain special doctor's appointments or medical treatments) for wanting a certain type of teaching schedule. Of course you want to (indeed *must*) attempt to accommodate these requests.

moderately fancy, then you may have to take the matter under advisement. For instance, if the faculty member wants a RAID unit, or some kind of fancy, high-speed color scanner—something that costs a few thousand dollars—then at the least you may have to consult your Executive Committee. You will need to determine whether the proposed use justifies the expense. Another approach is to determine whether several members of the faculty could make good use of this high-tech gimcrack, and then buy one for general use.

Some faculty members—especially the neophytes—may request help in preparing a grant proposal. You may be qualified to provide that help, or you could hook the tyro up with someone who can.

6.7.20 Endowed Lecture Series

My own math department is very fortunate these days because we have five different endowed lecture series. Each of these is a privately funded event, taking place once per year, that brings in a distinguished speaker for one or several days to give some talks.

Of course we have past Chairs to thank for raising the money to endow these different lecture series. Because they are *endowed*, they are a permanent feature of our mathematical life. We can depend on them year after year. We have one lecture series in geometry, one in analysis, one in algebra, one in probability, and one for undergraduates. People have a lot of fun with these events. It is a pleasure to pick the speakers. And we can offer an attractive honorarium and other perks, thus increasing the likelihood that the guests will accept our invitation.

Endowed lecture series are a significant component of the tapestry of mathematical life. They are something that everyone can rally around—not just the mathematical part but also the social activities pertaining thereto. And it doesn't take a great deal of money to endow a lecture series. About $40,000 will do it nicely. Donors like to give money for an activity of this sort, because it can and probably should be named after them. Each year you can give the Ingrid Toth Lecture Series on Algebraic Geometry, and you can invite the good Ms. Toth to the proceedings. Have her make a few remarks. Certainly create a fuss over her. Have her to the tea and to the dinner. This is the way that you build a relationship with a supporter of the math department, and curry future giving (which could and should be of a larger caliber). The initial rather modest gift that endows a lecture series could later grow into a more magnificent gift that endows a Chair.

If your department does not have any endowed lecture series, then perhaps you could spearhead an effort to raise the money for one. Talk to your Executive Committee about what needs to be done and how to do it. Give a call to your University Development Office and get something started.

6.7.21 Faculty Complaints

If some (faculty) member of the math department is unhappy about something then, likely as not, you as Chair will be the first to hear of it. Sometimes a colleague will break rank and go around you to the Dean. You hope that this won't happen. But it might. Then you are going to have a political situation to mop up. Let me deal with that elsewhere.

My view—when I was Chair—of faculty complaints is that I welcomed them. In the sense that I *wanted* to hear from my faculty, and I wanted the opportunity to exchange views. And it happens that I have a particular talent for helping to put these matters into perspective, and showing my interlocutor that this is something that we can deal with *together*. I could usually turn these exchanges into something productive, a catalyst for cooperation to solve a problem of mutual interest. This is a technique that you ought to consider cultivating in yourself.

Of course some faculty complaints can be downright nasty. Sexual harassment charges are no day at the beach. Allegations that a colleague has plagiarized, or misappropriated ideas, are ugly. With any situation like this you are going to need professional help. Your institution will have particular people in the administration who specialize in the legalities and the psychological issues attendant to such charges. Before you take any action whatever—and in fact *before you offer any substantive comment*—you should consult these good folks. They are, after all, there to help you.

There *are* fun aspects to being Chair, but handling faculty complaints is not one of them. Nonetheless, you can learn to turn these transactions into something you can build on, and in fact a way to strengthen your relations with your colleagues. You are like a union steward, looking after the best interests of the math department faculty. If you can convey that idea to your colleagues, you will all be better off.

In some instances you will be able to handle the faculty complaint math autonomously: There is a situation that needs to be rectified and you can in fact rectify it. In other instances, if the matter is something substantial (or if it will affect several people), then you may need to form a committee or task force to investigate the matter. In other cases (if it is a particularly sticky matter), you may have to consult the Dean for guidance. And, finally, there will be complaints that really don't have much merit. For those you will have to calm the complainant down and help him/her to see reason.

In all instances you should exercise patience and diplomacy and try to send the person away feeling good about the transaction. That is really your job here.

6.7.22   Faculty Discipline

It does not come up often—thank heaven—but sometimes a member of your faculty will need to be disciplined.

If the infraction is sexual harassment or something of that nature, then it is a legal situation, and higher powers will take over. You, the Chair, will play an ancillary role.

If the faculty member is guilty of financial misconduct then, again, you as Chair cannot really act alone. It is not so easy to engage in fraud as a member of the math department. You would really have to work at it. One possibility is that the individual borrowed from petty cash and failed to pay back the "loans." If that is the case, you may be able to just get the person to pay up, give him/her a good talking to, and then get back to business as usual. But I know of a case, for example, in which the Chair was in cahoots with the Administrative Assistant to file fraudulent travel reports and pocket the ghostly reimbursement checks. This clearly is thievery and fraud. They were caught and sent to jail.

In cases such as this last, the administration will have to become involved. And it is really not so pleasant, nor very pretty, to have to be the whistle-blower. If you become aware of an infraction of the sort just described, you would probably do well to find a colleague who will support you in the effort to bring the offenders to justice. And you might be smart to hire a personal attorney as well.

If, instead, the faculty member has failed to meet his/her classes, then you the Chair will be directly involved. You will have to document the dates and times and extent of the misdeeds. Was the faculty member routinely late, or did he/she actually miss class? How many times? And with what consequences for the students? Did this occur just in one semester—perhaps when the faculty member's spouse was hospitalized—or has it occurred repeatedly? Have you, the Chair, tried to discuss the matter with the faculty member, and with what effect? Is some improvement on the horizon?[30]

If you are put in the position of having to level threats against the uncooperative faculty member, then you will have to do so in concert with the Dean. The Dean has the power to actually take measures against the intransigent faculty member. You really do not (unless the Dean is behind you). You will find that the Dean has very little sympathy for faculty who do not acquit their teaching duties reliably and professionally. He/she will be quick to act if you bring in a case of dereliction of duty.[31]

Everyone makes mistakes, and everyone should be given a chance to rectify those mistakes. It is the recurring problems that must be dealt with decisively. If you are lucky, you can sniff out these problems early on and nip them in the bud before they become felonious. If not, then you will have a price to pay with your time and your effort and your peace of mind.

6.7.23　Prima Donnas and the Like

A prima donna is a female, solo opera singer. The prima donna is often a soprano, and is generally the star of the show. People will attend an opera because of who the prima donna is that evening. Because of the popularity factor for prima donnas, they are often confused with divas.

Maria Callas was a prima donna, and so was Beverly Sills. While Beverly was a kind and gracious lady, generous with all who worked with her, Maria was something of a contentious pain in the neck. Indeed, prima donnas are

---

[30]There was a famous case about thirty-five years ago of a tenured Professor at an important university in the Northeast who ran a publishing company in another city. He was absent a *lot* from his job. He generally got other faculty to cover his teaching for him, but they got tired of it after a while. Complaints were filed. And at some point the Chancellor issued this faculty member a forthwith: You will be in front of your class on thus and such a date, and continuously thereafter until the end of the semester, or you will be terminated. The faculty member said, "Oh, Chancellors never do that. I'm safe." He was terminated. And he never again held a faculty position.

[31]Another fairly well-publicized case concerned a Professor in Illinois who was the world's foremost expert on nude beaches. Really. He was very much in demand as a consultant (for expert testimony) whenever there was a court case anywhere in the country involving nude beaches. He would be flown in, put up in fancy hotels, and paid a handsome honorarium to provide his testimony. So naturally he was having a ball. But he missed *a lot* of classes. He worked in a department where teaching was the main thing; most of his colleagues really didn't have much of a scholarly life to speak of. And they were quite jealous of Mr. Nude Beach. They got tired of covering his classes for him, and they finally just turned him in. He was dismissed from his tenured position.

reputed to be vain, hot-tempered, and difficult.

The arena of mathematics has its fair share of prima donnas. They are usually *not* female—just because the profession is still dominated by males. The guy/gal who just got a paper into the *Annals*, or who just got an invitation to the International Congress, or who just won the Steele Prize, or just got a big grant, may be your local prima donna. It will be your privilege, and your burden as Chair, to deal with this person.

The prima donna mathematician is liable to have many demands. Surely he/she will want a very nice salary. And one of the nicest offices in the department. And perhaps more than the usual allotment of secretarial help. He/she may want a special travel fund. If this person really *is* a quite distinguished mathematician, then it is likely that many people will want to visit him/her. And you, the Chair, will be expected to pay for this travel.

Of course you want to take care of the accomplished people in your crew. The prima donna no doubt adds quite a lot of luster to your math department, and helps to make it famous. The Harvard department was famous in the 1950s because it had Oscar Zariski and Lars Ahlfors. Nobody remembers very many of the other people, although Mackey, Walsh, and Bott have their adherents. Fortunately Oscar and Lars were reasonable guys, and overall added to everyone else's pleasure in life.

The trouble, of course, is that your resources are limited. You have plenty of other individuals in your flock who work hard and accomplish a lot. They may not glow in the dark like your favorite prima donna, but they are quite deserving of emoluments and recognition. You will have to work with your Dean to do right by everyone. If you can arrange an endowed Chair Professorship for your prima donna—not at all an easy task—then that will solve your problem in one fell swoop. For then that particular faculty member is no longer on your budget—he/she is instead on the Dean's budget (because the Dean controls the Chair professorships). So the prima donna's raises, and all his/her perks, are under the Dean's purview. And you can use the limited raise money given to you each Spring to look after all your other faculty.

Many of your duties as Chair are a balancing act, and this item of handling your celebrity faculty is just one of them. There is always the danger that a fancy faculty member who feels that he/she is not being treated right will get an outside offer and leave (see the next section). You will have to cross that bridge when you come to it.

6.7.24 More on Outside Offers

One of the more challenging and interesting events that can occur on your watch is that one of your faculty will come to you with an outside offer (see also Section 4.19 for the faculty member's point of view). This means that he/she has a job offer from another university—presumably with a nice salary and other accoutrements to make it attractive.

You will have to examine your conscience to decide how to respond to an outside offer. You absolutely want to treat your colleagues with dignity and respect. The courteous thing to do is to make a competitive counteroffer. But one of the most frustrating things about being Chair is that there are lots of things that you want to do but you lack the power to do them math majorously. So you will have to go to the Dean.

The Dean always has his/her eye on the budget, and he/she will probably require some convincing. If the outside offer is from Harvard or Princeton, then you hold a lot of good cards. If instead the outside offer is from Little Sisters of the Swamp College, then you may have a problem on your hands. I know of more than one situation in which a Dean simply refused to respond to an outside offer. He said, "I guess that this person wants to go there for personal reasons. I am not going to make a counteroffer for an institution like that one." In a situation like this, you may have some discretionary funds in the department that you can use to sweeten your colleague's position a bit. Or perhaps you can adjust the teaching load down for a time. Or set up a travel fund. Or arrange for some nice raises for a few years. But, without the Dean's cooperation, an actual counteroffer may not be possible.

If the Dean *is* willing to play ball with you, then you need to have all the relevant information at your fingertips. Find out all the details of the offer, and all the perks. Find out what is special about this school, and why your colleague thinks he/she wants to go there. Find out about his/her spouse's situation. Is the competing school doing something special for the spouse? Can you respond to that? What about his/her kids?

There is only so much you can do, especially since you as Chair do not write the checks. In many instances—certainly not all—the faculty member with an outside offer is simply looking for some attention and some kudos. After all, it is a lot of trouble (and quite disruptive) to move. If you can make a civilized and attractive counteroffer, then you may have a fighting chance of getting him/her to stay. If, however, the outside offer is from the candidate's "dream institution", or in the candidate's "dream location", then you may be dead in the water no matter what you do.

As you work on responding to an outside offer for a colleague, you must also consider how the rest of the department is going to respond to the situation. Suppose that (with the Dean's acquiescence) you double this person's salary. How are the rest of the faculty going to feel about it? Many of them may say, "This guy/gal is no better than I am. It's just the luck of the draw that he/she got the outside offer. So my salary should be doubled as well." It is difficult to fault that reasoning. If the person in question is a highly valued and well-liked member of the department, then everyone will appreciate that a suitable counteroffer must be made. But you must suit the action to the circumstances and the context.

The fact is that excellent people are movable. And they will move. We must learn to come to grips with it. And then we hire more excellent people and life goes on.

6.7.25   The Library

The library is an essential part of scholarly life. In the best of all possible worlds, the math department will have its own library right in its own building.[32] And also have its own dedicated Math Librarian. The math collection could also be in the main library—with all the other books. That is OK, but slightly less convenient.

Mathematicians really need to have all the latest books and journals at their fingertips. Interlibrary loan is OK, but it causes a delay of from three days to three weeks. A good idea could be lost or go into hibernation during that period. Often, by the time the book finally arrives, you have forgotten why you ordered it.

Thus it is incumbent upon the Chair to look after the library, make sure it is well-funded, that the right books and journals get ordered, that the Math Librarian is an asset to the department. Typically the Chair will appoint a colleague—as part of his/her committee duties—to look after the library. But the Chair should keep his/her finger on the pulse of the situation. These days, with so many resources available electronically, and so many books and journals available at reduced prices through bundling, the library decisions

---

[32]At the Universite math majora de Madrid they brag that the math library is right there in the building. And indeed it is. But the building dates from the Fascist era in Spain—it is a huge edifice. In fact the math library is about a twenty-minute walk from the math offices.

are technical and complex. The Chair will do well to have a good relationship with the Math Librarian so that decisions can be made in the full flower of good information and knowledge of what all the choices are.

6.7.26 Computer Support and Technology for the Department

Any time the department develops a new need, then new funds are going to be required. And, absent private donations, you the Chair are going to have to get those funds from the Dean.

Thirty-five years ago, departments had to convince the Dean to buy or lease a big, fancy photocopy machine for the group. Prior to that, people had reproduced their exams and even their research papers with either ditto masters or mimeograph masters. This was quite a primitive technology, with roots in nineteenth-century methods of document reproduction. But it was all we had. The great thing about a ditto or a mimeograph machine is that it only cost a couple of hundred dollars, and it required very little maintenance. Also the cost per copy was very low. A photocopy machine is several orders of magnitude different in every category. Edwin Beckenbach, one of the grand old men of the UCLA math department when I knew him in the late 1970s, told me that the biggest fight he ever had with the math department when he was Chair occurred when he advocated that they get a motor-driven ditto machine to replace their hand-cranked one.[33] Even in the early 1980s when I was at Penn State the old-timers used a mimeograph machine to copy exams.

When I came to Washington University, as recently as 1986, the department was outsourcing its big photocopy jobs. It had no large-scale photocopy machine. Now it has two large, state-of-the-art machines right in the office. Photocopy machines require not only an initial outlay of money (although often they are leased) but also a maintenance and service contract. This is what the Dean calls an *entitlement*, which means that it's not a one-off expense. It is a continuing ding in his/her budget. So the Dean will think hard before saying "yes" to something like this.

A more recent fiscal development—especially for math departments—is computers. Now every faculty member[34] needs a computer on his/her desk,

---

[33]In those days math departments did little hiring, there were few people to tenure, and there was certainly never any discretionary money, so there was little else to fight about.

[34]The graduate students, too, need access to computers.

and certainly wants access to *e*-mail with a high-speed connection. Both of these cost money. I know people who work at small colleges where the computer support just isn't there; they buy their own Internet service with their own funds. At big universities we expect the institution to do it.

Any good-sized math department will want not only the equipment, but also someone to maintain the system. With fifty or more people networked, and printers and scanners, and a staff that is perhaps on a local area network (a LAN) so that they can share databases and spreadsheets, there are innumerable computer headaches that require constant attention. So you can just imagine the Chair going to the Dean and saying, "My department needs a full-time computer technician. These guys don't come cheap. I figure a salary of $80K per year plus benefits." The Dean would have myocardial infarction.

The Washington University physics department supports most of its computer people using soft money from grants. But a math department typically will not have that kind of money in grants. So the university has to pitch in. It is a battle to get this money, and to convince the Dean that computer maintenance people are now part of life and he/she will have to pony up. For many years a number of math departments maintained their computer systems with faculty volunteers—obviously not an attractive option.

In my own department we had to sneak our computer person in through the back door. We brought him in as a scholar from China on a grant that one of us had. He was a harmonic analyst who also happened to have some computer skills. Over time, his activities became more and more computer-oriented and less and less math-oriented. And his salary moved away from the grant and onto the department budget. Now he is a full-time, salaried member of the staff. We also have a regular budget item each year—of about $10,000—to buy new computer equipment for the faculty and staff. These are all new budget items that have come about in the last twenty years, and they had to be pried out of the Dean. But they are essential to the way that we now run our lives.[35]

---

[35]It also must be noted that, these days, a good deal of instruction is computer-based. For instance, many faculty use OnLine software for homework assignments, quizzes, and even exams. OnLine, self-grading homework systems—such as `Aleks`® from McGraw-Hill (a pay-as-you-go system) or `WeBWorK` (free to all users) from the University of Rochester—have become very popular. This creates a new set of computer demands. Many math departments now have extensive undergraduate computer labs.

6.7.27 Part-Time Faculty and Freeway Fliers

Most modern math departments do more lower division teaching than any other department in the College of Arts & Sciences. What with precalculus, calculus, discrete math, linear algebra, statistics, and differential equations, there is a great panoply of subject areas that need mathematics. And more and more disciplines—such as biology—are becoming mathematical. So the demand is increasing. At my own university, in a typical semester, the math department will offer about seventy-five courses while the chemistry department (which has more faculty) will offer about fifteen courses. Even with this growth in clientele, many math departments have shrunk in size in the past twenty-five years. Ohio State used to be the leader with about a hundred and ten faculty. Now it has about eighty-five. U. C. Berkeley and other major departments have also shrunk. But we still must cover our courses. We do *not* want the university to start pondering having Engineering teach its own calculus courses or physics teach its own differential equations courses. Don't laugh. These other departments feel that they are well-qualified to teach such courses (and like to have them taught "their way", which often means with little theory), and they would love to have the extra clientele.

Many parts of life consist in making peace among conflicting vectors. In the situation under discussion here, we want to do a responsible job of covering our teaching obligations. But we certainly do *not* want to raise faculty teaching loads. In the 1940s and 1950s, even people like André Weil taught three or more courses per semester. But that is no longer considered acceptable. Two courses per semester is considered reasonable, and many institutions have a lower teaching load than that.

If you take off your shoes and count, you realize that it simply is not possible to cover all the teaching obligations that a modern math department has using just the tenure-track faculty. Of course the instructors and postdocs contribute to the teaching effort. An important part of being a postdoc is getting some teaching experience. But that still won't cover the nut at many schools. Especially at big state schools, there is a *huge* clientele in precalculus and even pre-precalculus (today, in an effort at political correctness, these sorts of courses are given the moniker "developmental math"). So what is to be done?

The answer is to hire part-time faculty. These are people who are usually *not* under contract, and are hired course-by-course. The pay is not munificent—often between $2,500 and $4,000 per course. Usually the courses that these folks are given to teach will be subjects that the regular faculty don't want to teach—precalculus, statistics, freshman calculus, and the like. Part-time faculty, frankly, are often not treated very well. They usually do not have the full services of the staff; often they must do their own photocopying and other clerical tasks. They are certainly not provided with department-paid graders. They must do their own grading.

Some part-timers—the lucky ones—will get year-long contracts. They will teach six to eight courses during the year and receive a salary of $35,000 or so. These people do not have tenure, and do not have any of the perks that a tenure-track faculty member would have. They don't have their own offices; often they get by with just a cubicle, or a shared desk. They do not participate in faculty meetings, they do not direct graduate students, and they do not get regular raises.

Other part-timers are not so lucky. Known as *freeway fliers*, these folks will teach two courses at this school and one course at that school and three courses at some other school. If one is energetic and determined, then it is actually possible to make a living in this fashion. But it is draining and not very rewarding.

The harsh reality today is that many junior colleges have just a nucleus of permanent faculty and a huge collection of part-timers. With the budget that a typical junior college math department has, this is all that can be managed. As one goes up the food chain, one sees the proportions change. At Washington University we used to have a small collection of about four part-timers who did a really good job handling certain courses for us. But the Dean decided that a high-class place like Washington University should have all its courses taught by regular faculty. So we had to dismiss all our part-time faculty. This caused a major strain on our teaching resources, but that is how it is. It is ironic that the Dean ignores the fact that English uses an army of part-timers to teach freshman writing, and the French department uses a gaggle of part-timers to teach basic French. But math is treated differently. What else is new?

When I was Chair, it used to put quite a hit on my budget to pay our part-time faculty. For a time the Dean gave me money to help out. Then he stopped, as part of an effort to freeze them out. When I persisted in hiring part-time faculty, he finally just ordered me not to do it anymore.

The Princeton math department doesn't hire part-time faculty; the Rutgers math department—which is right up the road—definitely does. The Harvard math department doesn't hire part-time faculty; the University of Massachusetts math department—which is right up the (slightly longer) road—definitely does. This is the world we live in. It is an embarrassing but solid fact that often the part-time faculty are better teachers than the regular faculty. This is because in many cases they are young, committed, they are not trying to develop research careers, and they care. It gives the Dean pause for thought when he/she sees that someone who is getting \$4,000 to teach one course gets notably better teaching evaluations than someone who is getting a salary of \$130,000 per year.

6.7.28 Chair Professorships for the Department

One indicator of the Dean's regard for the math department is how many endowed Chair Professorships he/she is willing to give the group.[36] An endowed Chair is a special position for a distinguished scholar that entails

- a special salary

- a reduced teaching load, or special courses to teach

- a discretionary fund for travel, computer equipment, visitors, and the like

- a personal secretary

- invitations to special events

The endowed Chairs are really the royalty of the university. They are invited to special functions, a big fuss is made over them at every opportunity, and they are something to show off to visitors and dignitaries.

One great thing about an endowed Chair is that the holder of the Chair is taken off the departmental budget. He/she is paid from the chair's endowment, which is the Dean's funds. This means essentially that the department still gets the benefit of this person's presence, and this person's teaching, but doesn't pay for it. Obviously a good deal. Therefore, when the Chair is figuring raises, he/she can concentrate on everyone else; because the big shot

---

[36]Many public universities do not have endowed Chairs, or at least not many of them. At Harvard about half the math faculty hold Chair Professorships. In the UCLA math department there are just one or two. So my remarks in this Subsection may be primarily relevant to private universities in the United States.

is the Dean's problem. And all the big shot's perks are the Dean's problem also.

It's obviously a good thing for the math department to have several endowed Chairs, or at least to have about as many as chemistry or physics or biology has. At my own university mathematics has one endowed Chair while History has five. Chemistry has three. Physics has four. When, as Chair, I went to the Dean to discuss getting more endowed Chairs for math he would *always* give me the flip answer that, "Donors don't give money for math." And that was the end of that discussion.

It costs about $2 million to $4 million to endow a Chair, depending on how much the university is willing to kick in. A donor likes to give money for an endowed Chair because it is something quite visible that his/her name will be attached to: *The Joe Schlomokin Endowed Chair in Fractal Technology* is typical. It *is true*—consistent with my Dean's dismissive claim—that donors in recent years have concentrated on women's studies, the humanities, and the arts. Donations for Chairs in science are fewer. And chemistry and physics do seem to be more visible—in the public eye—than math. The big winner for endowed Chairs is medicine, largely because of the grateful patient syndrome. If the favorite nephew of some wealthy St. Louis scion is cured of cancer by the good doctors at Barnes Hospital, he/she will want to show appreciation in a big way. Endowing a Chair is a win-win situation for everyone: Mr. Gotrocks gets a Chair named after himself/herself, and the University gets a big plum. Unfortunately, nobody ever died of taking a calculus exam (although they may have *felt* as though they had).

I, as department Chair, endeavored to raise money for an endowed Chair for the department. I in fact worked with the University Development Office—and these are the real pros at drumming up funds. I was able to raise money for some smaller things, like an endowed lecture series and an undergraduate scholarship. But we never came close to getting a Chair.

If your department has an endowed Chair professor, and he/she announces that he/she will retire, then it is standard for people to start to discuss who next should get the Chair. This is wrong-headed. The Chair does *not* belong to the department. It belongs to the Dean, and it goes back into his/her pool. In case the original money was given specifically for an endowed Chair in math, then it is likely that the Department can hang onto the position when the incumbent retires. Most Chairs are more generic in nature, and they will go back into the Dean's pool. The math department will have to compete for them along with everyone else.

Sometimes a wealthy donor will give the university funds for an endowed Chair in some field that the University would rather stay away from. Not too many years ago some rather tainted Middle Eastern oil money was offered for an endowed Chair in Petroleum Studies. The donors had a hard time finding an American university that would accept the gratuity, but they finally did. Around the same time some Ross Perot type gave Washington University about $15 million to endow five Chairs in the study of fractals. The University took the money, and endowed the Chairs, but it interpreted "the study of fractals" rather loosely, and the Chair Professors are in many different departments doing many different things.

6.7.29 Retirement and Emeritus Professors

Not long ago the retirement laws in the United States changed. It used to be that everyone retired at age sixty-five. Then it became seventy. In the early 1990s, as a nod to the "gray panthers", the retirement age was raised to $+\infty$ for all but those in a few select professions. One of the select professions was the professoriate. These folks still had to retire at age seventy. Then, in 1994, the law changed one last time: Now the retirement age is $+\infty$ for everyone. This means that you never have to retire if you don't wish to.

Well, it's more complicated than that. Clearly we can't have math professors who are a bit dotty, and who no longer can calculate a derivative. How could such a person teach? So what we ought to do is give a competence exam—right? The trouble with that is you cannot give a competence test just to the old guys. You have to give it to everyone. That's the law. And the younger guys/gals certainly don't want to take such a test.

Some states now do administer competence tests to *everyone*. So every five years each of the distinguished faculty, as well as the geezers, must sit down and show that he/she can solve a first-order linear differential equation, or integrate by parts. Most states don't. And that's the point.

There are certain individuals whom you as Chair would like to get to retire. And they don't seem to want to do it. There are a variety of reasons for this. Some individuals have no other interests. Their life is coming to the math department, doing a little teaching, and hanging out with the gang. Some individuals think that they need the money. Some people are committed to a certain program, or a certain curriculum, and want to manage it. Some people are just stubborn.

There are lots of selfish reasons for you as Chair to want to get rid of these people. One is that they may no longer be the best teachers. You could probably do better by hiring some bright young guy/gal who represents an active research area and will inject new energy and life into the department. And if you can get this older person to retire, then you just might be able to convince the Dean to give you a new position and hire such a ray of light. And for much less money! So how is this done?

Sometimes the university will help. Just as an instance, the University of California has in recent years had two very attractive golden parachute programs. For people who had been in the saddle for twenty-five or thirty years, it was almost impossible not to retire—the financial incentives were just too strong. Plus one could still go back and teach a course now and then if one wished to do so—and be paid for it. And one could keep an office in the math building. What could be better?

One of the reasons that U.C. could institute such a program is that the University of California retirement fund is a separate financial instrument from the regular U.C. budget. And, in fact, it is quite well fixed these days. Unlike the Social Security system, it has plenty of excess cash, so it can afford to make allotments to get some of the old codgers off the books and into a happy retirement.

Not every school can make such an arrangement. Lots of college and university faculty have TIAA/CREF plus Social Security for a retirement package. That has much to recommend it, but it doesn't have the built-in latitude to create golden parachute programs. So then it is up to the Dean to do something. Our Dean will offer two years' full salary with no duties to anyone who will sign on the dotted line and agree to retire.

A few years ago we had a faculty member who had some severe medical disabilities. As a result he had a hard time teaching well. He still enjoyed working on his own, and doing mathematics. But the medications he was taking interfered with his hearing and his vision, so he could not be a good and effective teacher. But he didn't want to go on disability, in part because that arrangement was not financially very attractive. So I, as Chair, had to sit down and talk to him and his wife and convince him that we would all be better off if he retired. And, of course, I outlined things that I, with the Dean's approval, could do for him to make his life comfortable and rewarding and to ensure that he would still be considered an active member of the department. This actually worked, and he agreed to retire.

When one of your colleagues does retire, it is nice for everyone if you can treat him/her with some dignity and respect. The custom at Princeton, for example, is that when you retire they ask for your office key and wave goodbye. That's it. The only faculty member who was allowed to retain an office after retirement was Solomon Lefschetz, and he used to come in to work when he was well into his nineties. The reason that Salomon Bochner became Chair at Rice University after his retirement is that he felt no longer welcome in Princeton.

Many of us can do better than old Princeton.[37] Often the Chair can arrange for the retired professor to have an office—though it may be a shared office. He/she should be allowed access to a telephone, to have a library card, be able to use a computer with a high-speed Internet connection, and to ask for small favors from the staff. In some cases it is possible to arrange for the retired faculty member to teach occasionally—assuming that he/she wishes to do so—and even to be paid for it. It is very pleasant to be a retired faculty member, with no committee duties or other tiresome obligations, and still be able to offer a class from time to time. It makes one feel as though he/she is still part of the enterprise, and still contributing.

It can be a problem if a retired faculty member comes in with a one-hundred-page mathematical manuscript and wants it typed. These old guys don't know from TEX, and they remember the old days of manuscript typists. When this happened to me during my time as Chair, I—in an effort to be accommodating—had to use department funds to hire a student to type up the paper. It cost me about $500.

A standard courtesy is to grant a newly retired professor the status of *Emeritus.* This is sometimes done by a vote of the faculty and sometimes simply decided by the Dean. The "Emeritus" nomenclature is just a way to formalize the fact that the retired faculty member is still a member of the group and still has certain rights and privileges. An acquaintance of mine recently had a very unpleasant separation from his mathematics department. He was, in effect, forced to retire—and was banned from campus. It was pretty ugly, and there were a lot of lawyers involved. Part of the settlement is that everyone is under a gag order, so that the rest of us cannot determine what really went on there. Another part of the settlement is that this guy is *not* allowed to refer to himself as *Emeritus.*

---

[37]The tradition in the UCLA English department has been that, once a faculty member has been identified as "over the hill", then his/her office is moved into the basement of the English building. Thus the object of these affections obviously gets the message, and begins to ponder retirement.

6.7.30   Tenured Faculty Changing Departments

When I was an undergraduate, I was in the audience for a lecture by some big shot at U. C. Berkeley who announced during his talk that he had just decided to change departments—from mathematics to celestial mechanics. He allowed that he had done so "for epistemological reasons". I was flabbergasted. I could hardly imagine being a tenured faculty at any university—much less a prestigious place like U. C. Berkeley. That said, I couldn't imagine why one would want to change departments. What better place could there be than the good old math department? And how many universities have departments of celestial mechanics? What was that all about?

Well, people's interests change. If you as a mathematician begin a collaboration with a chemical engineer or a physicist, then the focus of your research could shift dramatically. It could, over a period of years, start to make sense for you to spend much more time in the allied department than in your home department. Does this justify a formal shift in your affiliation?

First I should say something about what the role of your department is in your life. After all, you are you. Your research is yours. Your teaching is yours. What difference does it make if you are even a member of the Etruscan philology department? The answer is this: **(a)** your tenure is to a particular department, **(b)** your teaching assignment comes from a particular department, **(c)** your committee duties are in a particular department, **(d)** your raises come from a particular department. So your departmental affiliation can shape your life and continue to affect it. In addition, your colleagues—the people with whom you will spend your day and share your thoughts—are the members of your department. These are your comrades-in-arms, and they influence the way that you see the world. You should not change affiliation lightly.

As a first step, if you feel a shift taking place, it could be arranged for you to have a courtesy appointment in the allied department. In practice, this means that you can check books out of their library, and perhaps offer a course in that other department. As a next step, with the cooperation of the two Chairs and the Dean, it can be designed that you will have a *joint appointment*. This is a real shift, and will impact your day-to-day life. For now you have two sets of teaching obligations, two sets of service obligations, and two sets of research obligations. And two Chairs have a say about your raises, your travel funds, and your other perks. You may find the joint appointment unsatisfactory—a case in which the whole is greater than the sum of its parts, but in a negative way—and then decide to make the full move. This will have to be done with the cooperation of the Dean and the two relevant Chairs. And the faculty may need to vote on it. It is liable to be complicated, but in the end may be a very positive change for you. Your original department Chair will have a lot of say in this lateral move, and also a vested interest. For if you move from the math department to the physics department, then what happens to your old position in math? The Chair would be foolish to just *assume* that the Dean will let him/her just fill it with someone of comparable rank. These days *all* positions are treated as new, and must be argued for from scratch.

6.7.31 Raising Money for Your Department

Most Chairs give little thought to the question of raising outside funds for their departments. Each year the Chair is given a budget, and he/she works to live within it. The Chair figures that the university has a Development Office, and those are the folks who know how to raise money. So let them do it.

This is a perfectly reasonable attitude, and a widely held one. But the Lord helps those who help themselves. I know one successful Chair at a big research university who managed to raise $3 million from department alumni and corporate donors in his first three years of office. Imagine having such a fund for one's discretionary use! The income from this principal would be at least $150,000 per year, and that is quite a discretionary fund to have.

The fact is—and professional fundraisers will tell you this—that there are lots of people out there who have money to give, and who want to give, and they just need to be educated. You need to convince those good folks that they would be happier giving the money to *you* than to somebody else. There are good books (see [BRA] or [FLA]) that will help you to learn how to do this. You need to make the donor part of the math department family, invite him/her to events, and make him/her feel that he/she is contributing to a dynamic, exciting, growing enterprise that enhances the American educational effort. I have known some of these donors personally, and they spoke glowingly of how their relationship with the university had enriched their lives (while they concomitantly enriched the university!).

There is a particular style and form that a letter soliciting donations should take. The books on this subject will tell you what type of stationery to use, what shape the paper, what kind of stamp to put on the envelope, and how to write the return address. The point is that you *don't* want the recipient to drop your letter in the trash bin; you want him/her to *read* it and *think* about it. There are particular tricks for making this happen. Once you have your foot in the door, then you must arrange to meet the donor and make your pitch. And you must follow through.

Getting people to donate money—especially big money—is not a one-shot operation. It may take six months to a year to develop the right sort of relationship and then to come around to making the money pitch. It is an art, and one that you will have to learn. But the potential payoff is enormous.

6.7.32   The Interface Between the Dean and the Faculty

Your Dean—perhaps he/she is the Dean of Arts & Sciences or the Dean of Science or maybe even the Dean of Mathematical Sciences—manages a good many departments. It is quite common for a Dean of Arts & Sciences to manage twenty-five or more departments. Yours is just one of them. The Dean is most likely not a mathematician (in fact odds are that the Dean is a lab scientist), and probably doesn't know much about mathematics. If the Dean is a good Dean, and a Dean who wants what is best, then he/she will be good at managing people and good at setting goals for the university. But he/she cannot possibly have detailed knowledge of every discipline.

It is easy to understand these concepts in the abstract but sometimes hard to get your arms around them in concrete situations. You will go to a meeting with the Dean thinking you are going to say to him/her, "We have lost certain faculty through retirement or attrition. We have distinct and severe needs in arithmetic algebraic geometry, string theory, nonlinear PDEs, and several complex variables." And you expect the Dean to know what you are talking about. Well, he/she won't. If you want to communicate effectively and well, then you have to speak to people in their own language. What is the Dean after? What are his/her goals for the College and for the Department? Whatever they may be, they are surely *not* formulated in the language of arithmetic algebraic geometry and string theory.

If you are inexperienced as an administrator, then you may find this situation quite baffling. *How do you approach this situation?* You must learn to make the points that are obvious to a mathematician—about specific subject areas of mathematics, about hot-breaking fields, about particularly strong individuals or new movements in the subject—but put them in a language that is meaningful to the Dean. You need to play up programmatic needs, new majors that are emerging, new job opportunities for graduate students, new directions in science overall. This may be quite foreign to you, and you will need to think hard about these issues and to develop new skills that will serve you well in "Dean-talk". Your Executive Committee can help here.

If you are lucky, then the Dean will have a beneficent view of the mathematics department. This will at least make him/her a priori receptive to whatever pitch you may be making on a given day. Unfortunately, at many schools the administration has a dim view of mathematics. We are perceived campus-wide as poor teachers, as scholars who are isolated and do not interact with other departments, as scientists who have small grants (or in many cases no grants at all!), and as strange ducks who are generally uncooperative. Mathematicians by nature and by training are *not* team players; rather, we are more like free spirits, following our own drummer and seeking our own truths. If the Dean wants to find a senior member of the chemistry faculty, he/she phones up the chemistry department and says, "I want to speak to Professor Czogolz." The answer he/she will get is most likely, "Professor Czogolz is in his lab. Let me get him for you." If, instead, the Dean phones the math department and says, "I want to speak to Professor Krantz," the answer that he/she gets may be, "Oh, Krantz is playing Ping-Pong with his graduate students" Or "Krantz has gone for a walk in the woods" Or "Krantz is taking his weekly hatha yoga class". This unfortunately makes no sense to the Dean. He/she doesn't understand that *my* lab takes a different form from a chemist's. There are different activities that spur my thinking. He may, sadly, conclude that we are a bunch of unreliable weirdos and flakes.[38]

Part of your job as Chair is to explain to the Dean—in language that he/she will understand—who we are and what we are about and what we are after. The fact of the matter is that we are serious and hardworking scholars, and we are generally creditable teachers. But the Dean may need some help to see these virtues clearly. Conversely, you will have to go back to the math department and explain to your colleagues how the Dean has responded to your well-thought-out proposal and what will happen next. This last task

---

[38]Most mathematicians glory in the well-documented story of Stephen Smale doing the work on the higher-dimensional Poincaré conjecture—that led to his Fields Medal!—while lying on the beaches of Rio. See [SMA].

is not always pleasant, as your colleagues may come (all too easily) to the conclusion that you have become an apologist for the administration.

Generally speaking, you will find it most effective to bring creative proposals to the Dean that are formulated in language, and with values, that he/she will appreciate. Of course, if you are clever, then good things for the math department will be woven into any pitch that you put to the Dean; but the overall gloss will be that this is something that is good for the College and good for the undergraduate students (or whatever segment of the university population you may choose to benefit on that day). The Dean will not respond well to a proposal that sounds like "Gimme, gimme, gimme". He/she wants to think that you are a team player, and that you are taking the broad view (just as he/she presumably is).

Deans like long-term planning. Rather than go to the Dean each year with a narrow, limited view of what hiring you would like to do *right now*, you will find it better received to have a five-year plan for hiring. Such a plan should take into account **(i)** upcoming retirements, **(ii)** impending tenure cases (which may or may not be successful), **(iii)** new scholarly developments, **(iv)** hot-breaking fields, **(v)** curricular changes, **(vi)** new opportunities for graduate students, **(vii)** long-term medical issues for certain individuals, **(viii)** disability issues. Most Deans are hard-liners about retirements. They know full well that a faculty member can, and indeed might, hang around on the payroll until he/she turns eighty. So if you say to the Dean, "Joe Dokes is turning sixty-five next year; we need to think about replacing him.", you will find that the Dean is singularly unsympathetic— *unless* you have a signed letter from Joe Dokes stipulating precisely when he intends to retire.

A very important point, and one that you will have to explain to your colleagues over and over again, is this. If Professor Beans retires, or Professor Jeans takes a job at Harvard, there is no guarantee that the department will get the position restored in order to replace the lost individual. In fact the Chair must argue for each new position as though it were from scratch. Nobody is "automatically" replaced. The one type of replacement that is supposed to be sacred is that if an Assistant Professor is terminated (i.e., not tenured), then you get to replace him/her. This is to ensure that a department will not be tempted to tenure a weak candidate just to ensure not losing the position. My own experience as Chair was that even that sacred trust was not universally honored.

Always prepare for meetings with the Dean. Deans have an uncanny ability to be on top of all sorts of information and statistics that you've never even heard of. No matter what proposal or idea you come up with, he/she can drown you in disinformation or misinformation or double-speak. You can fight back by always knowing in advance what the topic of any meeting with the Dean will be, and having the relevant information at your fingertips. I got to be pretty good at this, and I was often able to turn the table on the Dean with information or statistics that he/she had never seen.

If you are going to the Dean with a concrete proposal, you should have a version of it with you in writing (and in multiple copies!). That way you can leave him/her something to think about. If, by some luck, he/she approves it on the spot, then you can ask him/her to sign or initial the proposal right then and there. It is also a good habit to send the Dean a memo after your meeting summarizing what was said and what agreements might have been reached.

6.8 Miscellany of Being Chair

Many departments will have a formal document that comprises the Departmental Bylaws. It is possible that this document is widely promulgated and that the department expects you to run the show according to those rules. In my own department (as Chair) I was the only person who had a copy of these Bylaws. I thought they were pretty silly, and I ignored them. But you may not be able to get away with this, and you shouldn't try.

The department may also have a Mission Statement. This is different from the Bylaws, because it does not deal with details like how to run a faculty meeting or how to distribute raises. Instead, it deals with philosophical issues about goals in research, teaching, and service. It talks about the role of the faculty and the role of the Chair. It talks about the place of the department in the university at large.

Good universities have regular outside reviews of their departments (as well as regular accreditation visits by regional agencies). The way this works is that the Dean will solicit three or four distinguished scholars in the field (for us it would be mathematicians) to come to your department and spend a few days evaluating everything that you do. They will evaluate the effectiveness of the teaching program, the vitality of the research effort, the production of Ph.D.s, the ability of faculty to find jobs for their students, the contributions to departmental service, the relations with other departments on campus, and most every other facet of departmental life. The review committee will examine who in the department has grants, who gets outside invitations, who serves on editorial committees, who serves on national committees.

Usually the outside committee is composed of people from quite distinguished departments, and they take their task most seriously. Typically they will interview every faculty member and a good many of the graduate students and undergraduate majors. The outside committee will write a report that will be taken to heart by the administration, and is liable to stand as a template for the department's development for some time to come.

Of course the Chair cannot actually *shape* what the outside committee does or how it operates. But the Chair can have an effect. The Chair can have a lot to say about who is chosen to serve on this committee. And the Chair will schedule the committee's activities. When my department had an outside review about eight years ago, the busy committee members could only spend two days with us. And the Chair allowed their schedule to include half a day alone with the Dean of Engineering—a man who had a draconian agenda *against* the math department. This was obviously a big mistake.

If your department does have an outside review, this will probably be mandated and arranged by the Dean. I recommend that you as Chair form a task force to help you deal with the situation, see that the proper materials are assembled so that the committee will have the information that it needs (information about faculty research, recent Ph.D.s, undergraduate majors, and so forth), and to make sure that the committee talks to the right people and spends its time wisely. Meet with the Dean to plan the outside review; make sure that this is a joint effort, designed and formed to achieve mutually agreeable goals.

An outside committee report can be a great thing; it truly carries some weight with the Dean. It will have a decided effect on the department's future (in particular, on future hiring). Use what discretion you have as Chair to make sure it comes out right.

Throughout this book I have emphasized the importance of hiring to the department. It is something that everyone cares about, and a process in which everyone will participate. I once asked Saunders Mac Lane why the University of Chicago math department apparently lost ground in the 1950s. Prior to that, their department was arguably the preeminent one in the country; after that it was not. His answer was that, in the 1950s, he was Chair for eight years. And during that time the administration *did not allow him to do any hiring.*

I have mentioned before that, after you meet with the Dean, you should send him/her a memo summarizing what was said and what was agreed to. This is also good practice when you have a meeting with a colleague. You don't have to generate a memo if you spent a half hour with your local ring theorist jawing about trout fishing in America. But if you talked about his/her raise, or his/her teaching assignment, or something else of significance, then protect yourself with a memo. The memo is just insurance that everyone remembers the meeting in the same way.

Some Chairs take great comfort in knowing that they can resign, and they subliminally plan to use this as a cudgel when the going gets rough: If the Dean won't play ball, then threaten to resign; if the faculty is churlish and uncooperative, then threaten to resign. I personally think that this is wrong-headed. You signed on for this job, and you should have a considerable commitment to seeing it through. When the situation gets tough, you should dig in and try to make the most of the circumstance. Resigning should really be the action of last resort. If you do resign, it will weaken your standing—both with the Dean and with your colleagues—and it will weaken the department.

A Chair gives up some independence. You cannot travel as freely as you once did. You cannot necessarily take weekends off. You can no longer be the departmental hothead. You need to comport yourself at meetings—in fact all day long—with dignity and grace. You can no longer come to the department dressed in cutoffs and a tee-shirt fraught with lacunae (after all, you may have an impromptu meeting with the Dean, or with an important alumnus). You are the face of the math department, and you represent Math to the entire campus. So you should be someone that we can all be proud of.

6.9 The Duration of Your Term as Chair

Math departments like the idea that their Chair is a rotating position. So, typically, a term is three years or five years.

The trouble with a three-year term is that you spend the first year figuring out how the job works, and you spend the last year looking at the door. So you really only have one effective year in the job. Five years is probably a more productive period of time to spend in office.

Deans take a different view of these matters. I have worked at universities where the Dean viewed the departmental Chair position as essentially a lifetime job. So the Chair of astronomy, the Chair of chemistry, the Chair of physics were all the Dean's handpicked henchmen, and had been in office for as long as anyone could remember. The math department, as usual, was the gang of bohemians and insisted that its Chair be rotating. This was a thorn in the Dean's side, because he wanted department leaders who were his handpicked crew, people who would help him implement his policies, people he was comfortable working with. Put a slightly different way, the Dean finds it annoying to have to break in a new math department Chair every three or five years. He/she would strongly prefer to get a good and reliable person in there for a stretch of time. It just makes his/her life easier.

The Dean at one of my institutions had a College Executive Committee whose members were precisely these handpicked Chairs, and he used that committee to veto decisions and policies that he did not like. So a department would send up a tenure case, or a new curriculum, or a new set of policies, and it would sail through all the usual corridors of power, and then it would reach the hallowed Executive Committee. If the Dean didn't like this new creation, he would just tell his gang, and they would veto it. And that would be the end of it. That's university politics for you.

No matter what your Dean's paradigm for the Chair's term, I think that the smart thing to do when you agree to serve as Chair, and when you have your all-important meeting with the Dean in which you seal the deal and shake hands, is to say that you will serve an initial five years, after which you can both assess the situation and see how everyone feels. That way you, in effect, leave your options open, and build some flexibility into the situation.

6.10   Staying Alive While You Are Chair

Your life has many dimensions. Becoming Chair may cause you to emphasize some and minimize others of these. But you want to keep your personal life on an even keel. You want to keep your spouse/partner and children happy. You want to still have some diversions and sources of pleasure. And you want to keep your research program going.

This is difficult. I spoke to one outgoing Chair at a major research institution (not in math) who told me that, during her five-year term, she put in eighty hours per week, all on departmental business. She had no time for anything else. In addition, three of her colleagues attempted suicide during that period. Just imagine. This would drive almost anyone around the bend.

You must learn serious time management skills in order to maintain a balance in your life. In my first year as Chair, I frequently came to the department saying to myself, "The afternoon is wide open. Today I'll be able to do some math." And I ended up doing stuff for the department instead. You really must learn to *delegate tasks*. And also learn to compartmentalize tasks temporally. Something might come up at noon on the day when you planned to spend the afternoon doing research. Fine. Put the job in your *To Do* stack and do it tomorrow morning. It will keep. It is a good idea, if you can manage it, to set aside one or two afternoons per week for yourself. This is both good for your psyche and sense of self-worth, and also for your productivity.

It is not necessary that you personally do everything—just see that it is done, and done correctly. You have a good many colleagues who can help, and you have a staff who can carry a good part of the load. Even the administration and its staff can pitch in to make your burden a little easier in some contexts.

I have always taken it as axiomatic that a Chair teaches little or not at all. When I was Chair I always taught one course per year. The norm in my department had been double that. Many Chairs do no teaching, and have no trouble justifying this policy. Your Dean may have other ideas. As mentioned elsewhere, I frequently taught the teaching course for graduate students because (a) this did not require a great deal of preparation or effort and (b) I wanted to send a significant message to the students and everyone else that we think teaching is important. At my university, even the Dean teaches a course sometimes. Different institutions will be different.

6.11 A Second Term as Chair

If you are reasonably successful in your first term as Chair—this could be a three- or five-year period—then you may be asked to serve a second term. Or even a third term. There are upsides and downsides to this situation.

First, the good thing about your second term is that you have already figured out how to do the job. There is no "burning-in period". Second, you have a chance to implement policies and ideas that you didn't get around to in the first term. Third, you may have developed a good working relationship with the Dean and the department and can now hope to push it further and get more for the department. Last, you may take some satisfaction in doing things for the department and may now feel that you can do even more.

One bad feature of a second term—and I have seen this happen all too often—is that you can get cocky. I have seen a Chair wake up one morning early in his/her second term and say, "I've got this job figured out. I can do anything. These guys/gals in the department don't know what's going on. I run the show. I know how to play it with the Dean, how to sell a proposal. *I am the math department.*" You may think that I am making this up, but I am *not.*

One such Chair at a major state university ended up, fairly early in his second term, crafting a plan with the Dean to have a Teaching Faculty and a Research Faculty in the math department. This was a perfect example of putting a pitch to the Dean that would appeal to his/her baser instincts, but that also had something in it for the math department. For the Dean's greatest ambition was to push this math department into the top ten in the country. The Chair's ambition was to get more resources for the math department. Obviously a marriage made in heaven.

Well, the Chair and the Dean got this plan all doped out. Then the Chair called a faculty meeting. He stood confidently in front of his flock and announced, "OK, gang, we are going to have a Teaching Faculty and a Research Faculty. The Teaching Faculty will do all the work and the Research Faculty will get all the perks. Fair enough? OK—you, you, you, and you are the Teaching Faculty. And you, you, you, and you are the Research Faculty."

He just dropped it on them cold. And you can imagine how they reacted. They marched as one person over to the Dean's office and said—no, shouted—"What in the bloody heck is going on here? The Chair of the math department has just announced that we have two faculties with two different sets of privileges. This is worse than draconian. What do you have to say about this?" The Dean's reply was, "Huh. First I've ever heard of it."

There are several lessons to be learned here. Number one is that you cannot always trust the Dean. Number two is to always first run new policies by your Executive Committee and other trusted colleagues. Number three is to have a little humility. Pride goeth before the fall. This particular Chair was summarily dragged out of office with his fingernails in the carpet—in spite of the fact that his Chairpersonship had been, overall, quite successful. And it took him quite a while to recover, and to get his life back together. There were a number of his colleagues who would never speak to him again.

Interestingly, that selfsame department had a different Chair not too many years later who was much more politically savvy. Working with the same Dean, he was able to actually implement the "two faculties" policy, and it still stands today. But he was much more clever: He never announced it to the faculty—he just did it. And he was sure that the Dean was solidly behind him before he began. And his first move was to set up the necessary infrastructure and reward system. That Chair is now the Chancellor at another major university.

Another downside of second and third and more terms beyond that—not unrelated to the dangers of hubris that I've just described—is that you can get stale. In your first term as Chair you will really try, really listen to people, really exert yourself to be fair and balanced. After a while you will become jaded, you won't care so much, you will take it as more and more obvious that the only person in the building who really understands what's going on is yourself. You will do less and less for the department and more and more to feather your own nest and advance your own agendas. I do *not* mean to suggest that you will become dishonest; but you could become lazy. And certainly less effective.

I would say that, for most of us, two terms is plenty. After all, you have another life. You are a mathematician with scholarly interests and a research program. Why not get back to it? You might enjoy it.

6.12 Do You Want to Become a Dean?

Probably not. But it could be the right thing for the right person. Read on.

After you have been Chair for a while, you may feel that you are a changed person. You have developed a new skill set, you've learned how to manage people, and you can get things done. You've benefited your department a good deal, and now you want to take the next step and move up into the administration. That way you can have more influence and affect a good many more people's lives.

Even during my time as a mathematician, I have seen more and more colleagues follow this path. Robert Zimmer at the University of Chicago started out as department Chair. Now he is President of the University. Richard Herman is Chancellor at the University of Illinois in Urbana. Britt Kirwan was Chancellor at the University of Maryland, then at Ohio State, and now he is Chancellor of the entire University of Maryland system. A great many mathematicians have recently served as Deans and Provosts—

I'll just mention Tony Chan of UCLA, Robert Fefferman of the University of Chicago, Hugo Rossi at the University of Utah, Benedict Gross of Harvard, and Ron Douglas of SUNY Stony Brook.

It speaks well for our discipline that we (as a group) have become more worldly and can contribute to the running of the university. It is always good to have allies in the upper ranks. Of course it would be foolish to expect that, as soon as a mathematician becomes Dean or Provost or Chancellor, then all riches will be showered on the math department. Quite the contrary. The new officeholder will have to bend over backwards to establish his/her objectivity, and the person may start out actually being extremely tough on his/her old colleagues.

Another consideration is that it is relatively easy to become Chair. Few others want to do the job, and there is a sense that everyone should take a turn, so if you are willing you will probably get it. The Deanship is another situation. There is usually a significant pay boost, and you will have a lot more power. Lots of people want to be Dean—including people from outside the university. So there will be a lot more competition for the post.

Being Chair, realistically speaking, is a half-time job. If you use your time well, you can still manage to do some mathematics and keep your scholarly interests alive. You can go to conferences, enjoy your colleagues as you always have, and learn new things. Being Dean (or higher) is quite another story. Those are *big* jobs, often requiring that you work seven days a week. And you do *not* get summers off. It is really a major change in your life path, and you should think carefully before you embark down this road.

Part III

Looking Ahead

Chapter 7

Living Your Life

I'm always conscious of the fact that I am part of a profession that is 80% permanently unemployed. So, to be working in any sense is to be privileged.

Derek Jacobi (actor)

You spend so much time in your profession it ought to be something you love.

John H. Johnson (media mogul)

Incomprehensible jargon is the hallmark of a profession.

Kingman Brewster, Jr. (educator)

Throughout the human experience people have read history because they felt that it was a pleasure and that it was in some way instructive. The profession of professor of history has taken it in a very different direction.

Donald Kagan (historian)

To the extent that the judicial profession becomes the daily routine of deciding cases on the most secure precedents and the narrowest grounds available, the judicial mind atrophies and its perspective shrinks.

Irving R. Kaufman (justice)

A professional politician is a professionally dishonourable man. In order to get anywhere near high office he has to make so many compromises and submit to so many humiliations that he becomes indistinguishable from a streetwalker.

H. L. Mencken (journalist, literary critic)

7.1   Time Management

The older you get, the more you realize how valuable your time is. Money is nice, but you've probably got enough to get along. Possessions are OK, but after a while they get tiresome. By contrast, each hour that comes your way is the only one (in that particular slot) that you will ever get. When you are in the position of Professor, or perhaps some other mathematical professional, you will find that there are a great many demands on your time—more than seem reasonable. In today's world, we are all being asked to do more and more with fewer resources. The fact of the matter is that you are probably more than competent to perform any particular one of the tasks that is on your plate. What is difficult is to find the time to do *all of them* adequately and to your own (and everyone else's) satisfaction.

Definitely the *wrong* way to go about your work is to begin to think about Task A and then, after a few minutes, say to yourself, "Oh my God, I'd better have a look at Task B". And then after staring at B for a while, you jump to Task C. And so forth. What you are doing here is spending just enough time with each job to remind yourself what it is, and then tossing it aside and repeating the exercise with the next job. But you are not actually *doing* or *accomplishing* anything.

The syndrome that I am describing is perhaps best understood with a particular example. Consider the job of refereeing a paper for a journal. Here is what most mathematicians do—no fooling! He/she tosses the envelope into the work pile on the desk. After a time—days, weeks, or perhaps months— the envelope surfaces. The mathematician picks it up and has a look at it. He/she reads the title of the paper. Maybe he/she goes further and reads the abstract and the statement of the main theorem. And, of course, he/she checks the end of the paper to see whether his/her work is cited. Then he/she says, "I'll have to get around to refereeing this eventually." And then throws it back on the pile. *Just what has been accomplished here?*

The answer is: absolutely nothing. And the next iteration will be the same. Likely as not, on the tenth iteration, the mathematician will say, "I'm never going to get around to refereeing this paper. I should send it back to the journal." And then he/she throws it back on the pile. What a complete waste of time.

First, this scholar has wasted several hours just looking at the paper and not doing anything. In the end, he/she punted and just sent the paper back. It is perfectly all right to tell a journal that you have neither the time nor the expertise to referee the paper that has been sent you, but you should do so *right away*, while the envelope is still in your hand. *Get the thing off your desk!* If you *are* going to referee it, then make that decision right away and then sit down and do it. It will take you a couple of hours to do a proper job. But just set aside those couple of hours and get the job done.

And this last illustrates a key point of getting your work completed. On any given workday (which, presumably, has about eight hours in it), say to yourself, "Today I'm going to tackle jobs *A*, *B*, *C*. I shall allot two hours for the first, three hours for the second, and two hours for the third." That leaves one hour for slack that you can apply to any of the three tasks that requires extra time. Of course, if you have some experience in your profession, you will be pretty good at estimating how long any given task will take. So these two- and three-hour time slots will be meaningful and will get the job done.

What's great about this method is that, at the end of the day, you can look back and say, "Today I accomplished these three things." You can go get a well-deserved beer, or work out, or go home and be nice to your spouse. You've earned your keep for one more cycle of the sun. The alternative, to jump around from job to job and not accomplish anything, is both self-defeating and psychologically debilitating.

Another self-defeating habit, one that we all have, is to spend a good deal of the day standing around in the hall or sitting in the coffee room and complaining about how much work one has to do, or how overworked and unappreciated one is. One key to success in the academic game, or in any professional avocation, is to have the *Sitzfleisch* to knuckle down and actually *do the work* that needs to be done. Shun all distractions. Get your coffee cup and your pencils in place and *do the job*. Promise yourself a nice lunch with friends—*after* you have put in three or four hours performing the tasks that you've been assigned to do. Repeat this process in the afternoon. Of course we all have distractions like teaching and committee meetings. But the fact is that you probably teach only six or nine hours in a week, and you can't possibly have more than six to eight hours of committee meetings in a week (if you are Chair then this figure could be higher). That leaves twenty hours or more to do the other stuff. If you don't waste that time, you can actually get quite a lot accomplished.

And it is especially important when you are Chair to use your time wisely. You could easily spend twenty or more hours per week just listening to your colleagues kvetching. Of course you should make yourself available. Of course you should have an open-door policy. But, just as you do with your students, you should have certain times when you "receive company" and certain other times when your door is closed and you work. Any Chair should have a secretary who protects him/her from the great unwashed, and helps him/her to manage his/her time. It is also your prerogative as Chair to let it be known that you always spend Tuesday afternoons at home—with the phone off the hook if necessary. This can be your time to do research, or read, or ponder, or whatever it is that you need to do for yourself.

7.2   Publish or Perish

One of the hardest things you will ever do, and also one of the most important and rewarding, is to develop your own research program, to prove theorems, and to write papers and publish. In doing so, you will develop an international reputation, you will receive invitations from all over to give colloquia, to attend conferences, to be a distinguished visitor, and so forth. This is really a great life. You get to travel to exotic locales, you will have friends and colleagues (and, we hope, collaborators) everywhere, and you will have a professional network that bolsters your activities and increases your productivity. When, in 1942, Logan Wilson (in his book *The Academic Man: A Study in the Sociology of a Profession* [WIL]) coined the phrase "publish or perish", he could hardly have dreamed of what a prescient and meaningful utterance this was. If you want to advance in the profession, if you want to establish a reputation, if you want to get tenure, if you want to get a research grant, if you want to get speaking invitations, if you want to earn the respect and admiration of your colleagues (and your Dean!), then you had better publish. The alternative may not be to perish, but you may end up wishing you had perished.

Having a successful publishing program requires good time management— you can't prove theorems and write papers by snatching a twenty-minute time interval here and there. You definitely need large chunks of time so that you can internalize all the necessary ideas and *concentrate* on your mathematics. Otherwise you will never get anywhere. The task also requires considerable tenacity and determination and, in fact, also a dose of courage. For in order to put your ideas forth into the scholastic marketplace, you must believe in them, and you must be prepared for criticism and (occasionally) calumny. Your papers will sometimes be rejected, your grants will sometimes not be funded, your application to attend a conference (or to get funding to attend that conference) will not always be approved. Always remember: It's not how many times you fall down; it's how many times you get up. Managing your time well will put you more in control of things. You will less frequently be caught by surprise, and you will more often be able to make the best of the hand you are dealt and move your work and your career ahead.

If you are *very* senior, *very* accomplished, *very* independent, and *very* original, then you may be able to sit alone in your office and crank out great work. There are precious few of us who can do this. You need to see and communicate with other mathematicians. You need to feel the buzz of mathematics going on around you. You need to go to seminars, watch other mathematicians do their stuff, hear the words, and talk the talk. That is why you are in a math department or a mathematics research facility. That is why you have colleagues and students. It helps a good deal to be able to share your ideas, to bounce thoughts off of others, to get reactions to your conjectures and surmises.

There are few among us who can maintain a vigorous research program for forty years or more. One device, which Steve Smale of Berkeley and Ed Nelson of Princeton and Nick Varopoulos of Paris and other key people have pulled off with success and aplomb, is to change fields every several years. This will keep you fresh and young, and maintain your enthusiasm and your interest. Many people stop doing research because their field dries up. If you are a moving target, then this cannot happen to you.

When you are fresh out of graduate school, if you are lucky and are one of those who hits the ground running, then you will enjoy a very fertile and productive period of ten years or more. You will be operating on momentum from your graduate school experience, ideas from your thesis advisor, youthful vigor, and intensity of purpose. All of these attributes wane with time. Fields change and even expire. New ideas germinate and blossom, and you cannot keep up with all of them. You will wake up one day and realize that you are no longer one of the movers and shakers in your field. You will have to reinvent yourself. Not everyone has the resources or the energy or the opportunities to do this. So, for many people, the research life quiets down after a time.[1] I do not necessarily subscribe to the old saw that mathematics is a young man's game; that you are by definition dead when you turn forty.[2] In fact, a fifty-year-old mathematician has the advantage of maturity,

---

[1]It is different in a field such as literature, for instance. There you spend twenty or thirty years reading and developing your ideas. You actually get *better* as you grow older.

[2]Alexandre Grothendieck quit mathematics when he was forty; David Mumford, at

experience, and a broad perspective. He/she may have an extensive network of collaborators and colleagues with whom to discuss ideas. Dan Mostow proved his great theorem about rigidity of symmetric spaces when he was fairly far along. Andrew Wiles was forty years old when he (and Richard Taylor) proved Fermat's Last Theorem. There are other examples. So be of stout heart. The best may be yet to come.

But the other message here is that it is not the end of the line just because your research program is getting quieter. The mathematical life has many different dimensions, and there are lots of interesting things to do. You could write a book. You could get involved in curricular development. You could get involved in a national project for the betterment of the profession. Your could serve as Editor-in-Chief of the *Notices of the AMS*. A good general rule is to keep your options open.

Going to tea every day, or as often as you can, is a friendly and salubrious habit. It will help you to keep up with what is going on in your department, and it will also give you a chance for informal discussions about mathematics or whatever you like. It will give you a sense of the rhythm of the department, of who is doing what, whose research program is thriving and whose is stuck. It is refreshing and invigorating. It could lead to new collaborations or new projects. After tea you have another one or two hours—should you care to use them—to get some work done or think about some theorems or both.

7.3 Tenure and the Like

After I got promoted to full Professor, the Chair said to me, "Well, now it's just a long, slow slide into the grave." That's one way to look at it. I was in fact delighted to be at the pinnacle of my profession, and looked forward to a bright future. There is hardly any more secure and comforting job than a tenured Professorship at a good institution of higher learning in this country. You can just spend the next thirty years or so treading water—if that is what you want to do.

Personally I have found it most rewarding, and most invigorating, to seek to reinvent myself every five or so years. At least take on some new projects (a new kind of book, or a new kind of course, or a new position of

---

a comparable age, left pure math and moved into *very* applied pursuits. Each, however, managed to contribute monumental amounts to pure mathematics before shifting interests.

responsibility), buy a new house, or move to a new community. It is all too easy to fall into a rut and then just stay there.

Traditionally, the purpose of tenure was to free the Professor to pursue controversial truths, to teach daring ideas, and to invest long periods of time in his/her research without necessarily producing any results. As a mathematician, it is unlikely that you will be tempted to teach any ideas so daring that public attention will be attracted. You might decide to devote ten years to trying to prove the Riemann hypothesis, and you might experience a loss of productivity during that period. Don't worry; tenure will protect you.

But it has been said—with some justification—that the main purpose of tenure today is to protect you from your *colleagues*. As we have seen lately with Larry Summers[3] at Harvard or Ward Churchill[4] at the University of Colorado, university faculty can come under fierce attack from fellow faculty—to the point of having to resign positions. I am not necessarily defending either of these men—both took extreme risks—but telling of what they experienced.

Tenure does give you a certain peace of mind, and the cerebral tranquility, to concentrate on big ideas and lofty goals. It gives you the freedom and discretion to change fields, or change research direction, or change emphasis in your line of research. It gives you the time, the support, and the resources that you need in order to be a creative individual. Having tenure is a special privilege, and you should make the most of it.

If you were always worried whether you would have a job next year, or whether your salary might go abruptly down, it would be hard to concentrate on your work. If you do find a big, dangerous[5] idea that you want to

---

[3]Summers was the President of Harvard University. He got in trouble for stating, in a public speech, that innate differences between men and women might be one reason fewer women succeed in science and math careers. Summers also questioned how much of a role discrimination plays in the dearth of female professors in science and engineering at elite universities.

[4]Churchill wrote an essay that argued that American foreign policies provoked the attacks [on the World Trade Center Towers on 9/11]. He described the people working in the World Trade Center as "little Eichmanns".

[5]In mathematics, for example, it might be dangerous to develop a program to prove that the Riemann hypothesis is false. Because nobody believes that that is the case. Were you to do so, you might find that granting agencies are not receptive to your ideas. And that conference organizers do not want to hear what you have to say. And perhaps that journals do not want to publish your papers. When I was at Penn State I had a colleague

pursue, tenure will give you the freedom to do it. It can be argued that the universities pay us half of what we are worth but give us certain perks in exchange—like tenure and free summers.

In any event, this is the life we choose to live. It is stable and comforting, and it gives us the opportunity to develop our own ideas and to create a scholarly life for ourselves. It is a good life if we make the most of it.

So how do you get tenure? Do you hold your mouth a certain way? Is there a magic pill that you can take? Is it more a matter of attitude or of style?

It is safe to say that, among the 2,850 institutions of higher learning in this country, there may be 2,850 different ways to think about tenure. I work at a major research university, but it wasn't very long ago that tenure decisions there were pretty much decided by a (brief) phone call between the Chair and the Dean. Today, for legal reasons as well as others, tenure procedures at various types of institutions are becoming structured and uniform. At least they have several properties in common. I shall describe many of them here, and also indicate some of the differences that might come up.

First, most every institution of higher learning has a so-called *Tenure Document*. You will not find this item widely disseminated around campus. It is not dropped in bundles from low-flying planes. But it sits in the Chancellor's office and the Dean's office, and you can get a copy if you need one. This document charts out just what tenure means *at this institution*, how you get it, and how you can lose it. It is a legal document that holds great portent for your life. So you would be well to acquaint yourself with it.

One thing that the *Tenure Document* will say is that a tenure case is judged on three factors: **(i)** teaching, **(ii)** research, and **(iii)** service. Sometimes there will be a veiled implication that these three factors are weighted equally. Other times not. I can tell you with some certainty that they are *not* weighted equally. They all count, and if you are dreadfully deficient in one category then you are likely not to get tenure at your institution. But different institutions will weight them differently.

At a research university—Princeton or Harvard or Berkeley or Chicago or the like—research is the main thing. Your tenure dossier must demonstrate

---

who had a well-developed program to show that the Poincaré conjecture was false. And he was going to prove this result *not* by providing a counterexample, but instead by using logic. His approach was quite original, and at first it garnered some attention. After a while people got tired of hearing it, and he was shunned. Finally some experts sat down with him and explained why this quest was doomed to failure.

beyond any doubt that you have a major research program, supported by ten to twenty published papers and strong letters of recommendation (about six to ten) from outside experts, in a particular field of mathematics. The papers should be published in independent, peer-reviewed journals of good repute. Deans are well aware of the pecking order among journals. A savvy Dean will know that a paper in the *Annals* counts a lot more than a paper in *Mathematicae Portugalae*. Deans also know (see Section 5.4) that a good letter of recommendation will have a number of specific components that show that you deserve tenure. These will include:

- A reasonably detailed discussion of your research, why it is significant, and how it compares with the research of others at your level.

- Explicit binary comparisons of you (the candidate) with other mathematicians at the same level. For example, if one of your letters says, "Benjamin Disraeli just got tenure at Johns Hopkins and this candidate is much more talented", then that will count for a lot.

- A description of your teaching skills (if the writer has this information).

- A discussion of your activities in the profession, your collaborations, the talks you give at conferences, your propensity to communicate mathematics and to work well with others. If the letterwriter can say anything about your service activities, then of course that information is useful.

A good letter of recommendation will be two to three pages long, and will be *very specific* about your attributes. It will make a very positive and very *explicit* recommendation for tenure. It will leave nothing to the imagination.

Your teaching must also be evaluated objectively and in detail. There are a variety of ways to do this. The Chair might send some professors to your classes to observe your teaching. They will write a report, detailing your good points and your not-so-good. The Chair will assess your written teaching evaluations from students and write a précis of what they say. The Chair will poll other faculty to see what they know about your teaching. Sometimes the Chair will solicit letters from some of your former students. The teaching aspect of the dossier must be concrete and full of real information about what you can do and how you do it. It must not sound like a bunch of fluff. The Dean needs to know that you are someone who will make worthwhile contributions to the teaching enterprise for the next forty years.

Finally there is your service. This, too, is taken seriously, but it is not on the same level of gravitas as the first two criteria. There needs to be documentation that you have served on committees—both at the department and the University level—and that you have done your homework and acquitted yourself honorably. If you have *volunteered* for service activities, certainly that will reflect well on your attributes as a departmental citizen. If you have done undergraduate advising, or directed Ph.D. students, that will be to your credit.

What you must do, as an Assistant Professor, to earn tenure is to work hard at all three of these desiderata. As noted previously, different schools will weight these factors differently. A research institution will count the research at least 75% and the remainder for the other two aspects. A primarily teaching institution will count teaching about 60%, service about 30%, and will want to know that you have a research profile but will not be too concerned about the details. A comprehensive school (see [KRA3] for the word about these schools—which are most of the schools in this country) will have standards that are somewhere in between.[6]

As I have said, a frontline research institution will write to prominent professors at Berkeley, Harvard, Paris, Göttingen, and elsewhere to get the straight poop on your research, and how it ranks in the field. But if you teach at Southwest Missouri State (recently renamed Missouri State University), then it is unlikely that such an exercise would be meaningful. You have probably done some research, but you don't work in a department whose main emphasis is research. You have many teaching and service obligations, and you simply don't have the time to develop a high-level research program. And you are not expected to. It would make absolutely no sense for your Chair to write out to Paris or Harvard and ask for an assessment of your work, and he/she will not do so. Often tenure cases at primarily teaching schools are decided largely inhouse. At one pretty good school where a friend of mine teaches, he was asked simply to fill out a form explaining to the Dean why he deserved tenure. And he got it!

You yourself have very little control over the *tenure process itself*. You may be invited to suggest names of letter writers, or of people whom you would prefer not write letters. You may be asked to contribute information about your teaching or your service. But, for the most part, the spaghetti

---

[6]One math department at a comprehensive school of my acquaintance recently had a faculty vote to the effect that teaching will count 50%, scholarship 25%, and service 25%.

factory takes over and processes your tenure case. You simply must wait for the outcome. And you will be given few if any clues along the way.

7.4 How to Be a Tenured Faculty Member

I must say that, once I got tenure, my view of the world—and of life in general—changed distinctly. Now I was a vested member of one of the oldest and most dignified professions on earth. Now I was set for life. Now I could do—at least academically—just whatever I wanted, and for as long as I wanted, with whatever results I might care to produce. I was now one of the *haves*, whereas formerly (as a student and definitely as an Assistant Professor) I was certainly one of the *have not*s.

That is a selfish rendition of the picture, but nonetheless accurate. There is also a more charitable side. Now that you are tenured, you have the leisure and perhaps the power and wherewithal to effect some positive change on your surroundings. How can you make the overall level of teaching in the department better? How can you improve the quality of the mathematics curriculum? Can you develop collaborations with your colleagues? How can you make the department a more proactive unit that works effectively to achieve larger goals (than isolated individuals can achieve)? How can you create a dialogue between pure and applied mathematics? How can you effect some synergy between mathematics and engineering, or between mathematics and physics?

There is no easy answer to any of these questions. If there were, then I would bottle it and sell it. But this is why you have tenure. Now you can devote some time to one or more of these issues (presumably in collaboration with a friend and colleague) and make a memorable difference.

Another path to take—and this could be a *very* important one—is to pick a *really* big research project and attempt to make some headway on it. This would be the kind of problem that would be too daunting for an Assistant Professor[7]—the sort of thing on which you could easily spend six years and

---

[7]When I was Chairman of the Dean's promotion and tenure committee at Penn State, we had a case of a chemist who worked on synthesizing organic molecules. It was a zero-one game: You set your sights on synthesizing molecule XYZ. If you succeeded, then you were a big success and you got tenure. If you did not succeed, then there was no cigar. Well, this guy succeeded—right after the letters for his tenure case were solicited. So nobody writing the letters knew anything about his important molecular synthesis. Some

then have nothing to show (which would, of course, lead to a bad end for an Assistant Professor). Realistically, you would probably be wisest to both formulate and attack a goal of this magnitude in collaboration with a trusted colleague. It is all too easy to choose a project that is far too large for one person and then to get smothered by it.

One of the keys to success in the academic game is to choose problems that are the right size, that you can make a dent in, and that you can use to establish your reputation. If you choose, for example, to work on the Riemann hypothesis then you would be rather foolish to anticipate that you were actually going to solve it. Every major number theorist for the past 150 years—and many a fine analyst, too—has banged his/her head against this one. And nobody has come close. But you *can* hope to shed some light, to create a new approach, to suggest some new connections. This is what you must dope out before you actually engage in your multiyear quest. You can't just go in blindly and start swinging your fists around. It is impossible to succeed at anything unless you know in advance just what it is you are trying to accomplish.

7.5   What Happens If You Don't Get Tenure?

Most mathematicians are not cut out to handle failure well. Most of us have spent our lives being the best and the brightest and the smartest at everything that we did, and we sailed from one triumph to the next. This included getting into a great undergraduate institution and a fine graduate program.

But going up for tenure is a whole different ball game. Now you are competing—at least in theory—against all the best young people in your field. Why should your university tenure you if there is a bright young guy/gal at Chicago who is better and wants to come? This is the battle of a lifetime, and the stakes are high. It is like offering to sacrifice your Queen in chess: if it works, you win the whole enchilada; but if it fails, then you are cooked.

Getting turned down for tenure is a huge blow, and often it comes unexpectedly. This happens because frequently we are not getting good counsel

---

of the letter writers found out about this young fellow's coup *after* having submitted their letters. So then they sent in a second letter correcting the first one. You can imagine what a mess this case turned out to be. In the end the poor candidate was *not* tenured.

from our senior mentors; or it could be because we got the counsel but we didn't listen. In the best of all possible worlds, no Assistant Professor would go up for tenure unless he/she was virtually a sure shot for success. Anything less and either the Chair or the mentors would just tell the candidate, "Look, kid. This isn't going to work. We'll give you a year to pack your bags, and you'd better use it to find another line of work." Unfortunately, most math departments don't have their act together sufficiently to pull this off. Usually, if a candidate wants to go up for tenure, he/she goes up for tenure. And let the cards fall where they may.

Being turned down for tenure is not necessarily the end of the line. Today there are four famous Professors at Harvard and Princeton who were turned down for tenure at their first jobs. In two of these cases the failure was because the candidate could not get along with his senior mentor; in one case it was because the candidate could not teach; in the fourth case it was because the candidate was perceived not to be good enough. But these guys fought on, and they succeeded. They are now all world-class mathematicians. I myself was turned down for tenure at UCLA. It was not very pretty, but I was fortunate that a lot of good people stood up for me and helped me to land on my feet. And the experience made me stronger.

If, indeed, you *are* turned down for tenure, then you will want to explore what your options are. You may be able to get another academic job, but it will be at a lesser institution. Then you can fight your way back up the ladder. Or you might want to consider a job in industry, or in a research lab at a medical school, or with the government. The Social Security Administration, just as an instance, employs many mathematicians. The National Aeronautics and Space Administration (NASA) employs a number of mathematicians. The National Security Agency (NSA) employs more Ph.D. mathematicians than any other company or organization in the world. It is certainly possible to work in the private sector for a while and then go back to academics—if that is what appeals.

I know one mathematician who didn't get tenure and now runs a woodworking shop. Another runs a metal machine shop. Yet another is a TeX consultant. Another has started his own publishing house. It is a big world out there, and you are a talented person. There are a great many things that you can do. Being a professor is nice, but it's *not* the only thing.

7.6   If Tenure Doesn't Make You Happy?

There are those of us who will be unhappy no matter where we are or what we are doing. There are others who always seek new challenges. Still more will find the life of a professor to be too sedentary and not filled with enough challenges.

One of my Ph.D. students got tenure at a pretty good place but quit to start his own Internet company. The fact that his wife was unhappy where they were was a factor here, but it was not the only factor. Another of my Ph.D. students now develops software in Silicon Valley. He was a *very* talented mathematician and could have had a fine academic career. But, again, personal factors influenced his thinking.

Three of my former Ph.D. students now work at the National Security Agency (NSA). One of these had several academic offers, but did not want to move away from his family. Another just couldn't get an academic offer. The third—a *very* talented teacher and good researcher—decided a priori that she did not want an academic job. NSA offered a high starting salary and the opportunity to live in Washington, so she took it. And likes it very much.

Washington University had a very bright Ph.D. student finish a couple of years ago. He wrote a strong thesis, but through bad luck could not get an academic job. He ended up accepting a position in the financial sector in New York City. His starting salary was close to $200K.

Another friend was virtually a shoo-in for tenure at a *very good* place, but he psyched himself out. He convinced himself that anything he worked on—for a year or two—his thesis advisor could do in five minutes. So why should he waste his time banging his head against the wall? He went off to start his own publishing house. Yet another friend got a good job, with tenure, at a fine school in Colorado. She got sick of the academic politics, and quit to go to work for AT&T. She met her now husband there and is very happy with her life. One acquaintance of mine was put up for promotion to full Professor at his (rather good) institution. He received some negative letters, and the case did not go through. He subsequently quit his tenured job so that he could pursue his avocation of playing the guitar.

Adult life in this country consists of making choices—it is probably different in Afghanistan—and if one is lucky he/she makes more good ones than bad ones. Remaining a tenured professor all your life just may not be for you. It can become boring and repetitive and, indeed, cloying. Especially living in a small college town could drive you (or your spouse) bats. The candidate described at the end of the preceding paragraph almost certainly could have been promoted next time around—he just needed to clean up his act a bit. But he is probably much happier doing what he is doing now.

As you go through graduate school you will be indoctrinated—as we all are—with the idea that there is only one life worth living and that is to be a clone of your thesis advisor. That's an OK way to look at the world, but it is extremely limited. And it leaves you few choices if the academic life doesn't work out. Better to acquaint yourself with what *all* the choices are. That way you can make a more informed decision when the time comes, and do what is best for you and your family.

7.7   How to Keep Your Teaching Alive and Vital

I started teaching at the university when I was twenty-three years old. I turned out to have a knack for it, but I had a real advantage in being so young. I looked just like a student, I dressed like a student, and the kids automatically wanted to like me. They laughed at my jokes, they respected what I had to say, and for the most part my teaching was successful.

I will say that being a young instructor had its downside. Students find younger faculty to be more approachable, and I experienced an unreasonable number of tearful episodes over grades and other issues. I was approached about matters that a more senior faculty member would never hear about.

Now I am more like a grizzled, old veteran. The students still like me and respect me, but they perceive me to be more austere. I never have tearful episodes over grades anymore. Students do not approach me with outrageous or unseemly requests. Perhaps this is an improvement. But the flip side of all this is that I am no longer anything like a student. I don't talk their talk, and I don't walk their walk. I haven't seen the movies they've seen, I don't listen to the music they listen to. I read books and they don't. In sum, we have a lot less in common.

When you are older, it is perhaps a bit harder to connect with the students. You would like to have a good rapport with your classes, as it makes communication that much more effective. But you are going to have to try harder to make this happen. The students may not find your jokes so funny anymore. The students may find your remarks about campus life to be crusty and ill-humored. In short—and I am sorry to say this—the students may think you are a bore.

Some people decide that this is just the nature of life, and they slog through the teaching experience with ever-diminishing enthusiasm. They feel as though they are teaching alien beings, ones with whom meaningful communication is all but impossible. But these faculty have tenure, they know how to teach (at least in principle), and they just stand up there and do it.

There are ways to cut through this predicament. I have one friend who brings the newspaper to class with him each day. He picks out an article that he thinks will be a touchstone of common interest, and he discusses it with the class; or makes witty remarks; or asks for their witty remarks. This turns out to be a great class habit, and the students really look forward to it. Your class is like your (extended) family, and you can use devices, that are similar to ones you might use around the dinner table with your kids, to keep things lively in your class. Corny, self-deprecating jokes can work wonders. Especially jokes (or habits) that become like a daily ritual.

One of the more successful teaching heroes these days is Ole Hald of U. C. Berkeley. He uses the following device with great success. Imagine that he is teaching the chain rule to a large lecture class. The traditional way to do it is to present the idea and then give five examples. That will fill up an hour, and it is a perfectly reasonable way to proceed. But Ole has a better idea. He does two examples, and then he has the students do one right there in class. It is a short example that they can do in a few minutes. He makes a big ceremony out of it—there are assistants who hand out the examples on little slips of paper. When the students are finished with their task, Ole makes sure that they understand what the right answer is and why. He assesses how many people got it right. Then *he* does another example. And he finishes with the students doing another example.

What a great idea! The same amount of material is covered, but now the students are *empowered*. It is one thing for them to watch *you* do the stuff. But then they are likely to go home and—three days later when they try it—they will have forgotten everything. With Hald's technique they have actually confirmed for themselves on the spot that *they can do the chain rule*. Ole's students leave the room in an upbeat frame of mind, convinced that they have truly learned—and begun to master—something new. This is a real triumph: a simple idea that works like a charm.

If you give some genuine thought to your teaching—and you should do this every semester—then you will produce your own devices for getting in touch with your class. You can present vignettes from the history of mathematics. You can tell amusing anecdotes about mathematicians (see [KRA7], [KRA8]). I got one class so that they just loved Norbert Wiener stories. Students used to stop me when I walked across campus and beg for a Norbert Wiener story. You can show the students interesting snippets of mathematics that are related to but not directly a part of the course. Students should understand from the get-go that these "will not be on the exam". They are presented for everyone's cultural edification, and for just a little diversion.

In the 1960s I was an admirer of David Harris, who was the leader of the antidraft movement. A particular quality that he had was that he could talk to *anyone*. He could not only pitch the anti-draft philosophy to radical students; he could also pitch it to farmers, and Methodists, and Chinese immigrants. He could figure out where people lived and talk to them in language that they could understand. This is what you must learn to do. It is part of teaching effectively and well, part of your craft, and you would do well to cultivate it.

7.8 Promotion Through the Ranks

When I was a graduate student at Princeton and I attended a memorial service for Steenrod or some other eminent mathematician, I always marveled during the recitation of the person's career, especially the part where the speaker said, "He worked his way through the ranks and became a full Professor in 1963." Or something like that. It all sounds so routine. In point of fact, for many of us, it can be quite an ordeal to get tenure and another ordeal altogether to be promoted to full Professor. Different schools are different. The attitude at Princeton is that once you get tenure you should expeditiously be made a full Professor. The sooner the better and let's be done with it. At Harvard they have essentially eliminated the rank of Associate Professor. Their view is that if you are good enough to be tenured at Harvard, then you should be a full Professor. No need to shilly-shally around.

Other schools are much more lockstep in their approach to the matter. At some universities it is etched in stone that an Associate Professor must wait eight years, or ten years, or some other predetermined length of time, to be put up for full Professor. Never mind the merits of the case; everyone else has had to wait ten years, and this new guy/gal will have to wait ten years also. At other schools a candidate can be put up for promotion whenever he/she and the department feel that the case is ready.

At some schools the process of considering a candidate for tenure and the process of considering a candidate for promotion to Associate Professor are separate operations. Generally, the second comes before the first. At most schools the two elevations take place at the same time. You will learn to deal with the paradigm that is in place at your institution.

There is a certain delicacy to the matter of promotion to full Professor. If you are put up too soon for promotion and the Dean slaps you down, then it is demoralizing. And you will have to wait another three years or more to be put up again. If the Chair has a good relationship with the Dean, then he/she can get a preliminary reading on a case and he/she will know whether or not it is going to fly. Otherwise it is just a good bet. The Chair makes the best case he/she can make, and does his/her utmost to push it through the system. When he/she collects outside letters, they are sometimes disappointing, and do not support the case that the Chair wants to make.

Of course the Chair wants to be supportive of his/her faculty. If a particular individual feels passionately that it is time for him/her to be made a full Professor, then the Chair wants to help. But sometimes the case is just not there. The letters will not support the claim. The teaching and service are just not up to par. The publication record is lacking. Then the Chair is faced with the distasteful task of telling this candidate that he/she would be wise to wait; it would weaken the department and weaken the candidate's case to put him/her up at this time. This is one of the tougher aspects of being Chair.

Most every department has certain individuals who have been pegged as "lifetime Associate Professors". These are people whose research program has gone into stasis, who are not making any special contributions to the teaching program or to the curriculum, or whose service record is nothing spectacular. Generally speaking, these are good and likeable people. Their overall contribution to the quality of life is positive rather than negative. But they simply do not meet the objective criteria for promotion. As a result, these individuals often feel beaten down, they have the hope of a better life wrung out of them, they have no expectations for bettering themselves in this lifetime.

The Chair must deal with these cases. If he/she can (and this is *very hard*), the Chair should try to work with the individual to strengthen his/her case and ultimately put the person up for promotion. He/she should consult with the Dean and work with him/her to decide what needs to be done in each instance. Not everyone is ready to bring a moribund research program back to life. But there are other useful and worthwhile paths that a scholar may pursue; there are other kinds of publishing, there is curricular work, there are teaching initiatives. One Associate Professor at a big state university wrote an important book and got promoted. Another created an outreach program for high school students; this had the additional benefit that it attracted more top students to the institution, and strengthened the math major. That was enough for promotion. Still another Associate Professor (this time at a private school) became the key person in the department for the undergraduate program, and served as Vice-Chair of Undergraduate Studies for many years; the Dean was happy to promote him.

One can dynamically demonstrate his/her special talents as an administrator by taking a person who has been written off as deadwood and bringing that person back to life. The Chair can consider that to be one of his/her true accomplishments.

7.9 Striking a Balance

We would all like to think of ourselves as hotshot mathematical researchers or scholars—not just now, not just for the first several years of our careers, but actually for the duration. We would like to think that the papers or books that we write when we are seventy years old are just as exciting and vital as the papers that we wrote when we were thirty. Probably this is not the case, and there is external evidence to prove it: **(a)** You are not enjoying the speaking engagements that you once did; **(b)** You no longer have a research grant; **(c)** Your presence at important conferences does not seem to be as essential as it once was; **(d)** You no longer have vibrant collaborators clamoring to work with you; **(e)** You no longer have any graduate students.

Be of stout heart. We all grow older, and as we do we become less a part of the mainstream. It is nothing to bemoan, for you can develop new interests. You can make new contributions. You can garner new friends.

Unfortunately, mathematicians often are not built this way. We are trained from the get-go to be excellent, to be at the top of our fields, to be on the cutting edge. We have spent our entire lives being the best at what we do, and being praised for it. But it is the ordinary course of things that the young geniuses of today become the senior mentors of tomorrow, and new young geniuses take their places. There are an exceptional few among us who really do maintain the old fire-in-the-guts well into their seventies. But for most of us some adjustments are in order.

As you move into your fifties, it will start to seem reasonable to serve a term as Chair, and you should consider the option seriously. Or at least serve as Vice-Chair for Graduate Studies. Some universities have rotating Deanships: The term is three years and then you are out. This is to give a variety of faculty the experience of being an administrator, and also to keep the position fresh with a flow of new blood and new ideas. That may be something that you wish to try.

Some senior mathematicians that I know keep their hands in with research, but give the bulk of their time to developing fruitful relationships with local high schools. Especially when people have kids in school, they develop a particular passion for the quality of public education. The Math Wars in California (see the essay by David Klein in [KRA1]) were largely fueled by Stanford and Berkeley math professors. The long-running brouhaha over the NCTM (National Council of Teachers of Mathematics) Standards for Math Education was largely fought by college and university faculty. And these are not minor dalliances. They are fundamental public issues that affect all of society.

I have one friend, an outstanding geometer, who now devotes all his time to high school teacher training. He runs workshops every summer to bring high school math teachers up to speed in geometry and other key subjects. I have another who has become the go-to person for school textbooks. There are so many innovative and unusual approaches these days, such as the CORE PLUS textbook series, that it is good to have someone who has read all the books and can compare them critically. This particular individual is a member of the National Academy of Sciences and a very distinguished mathematician. But it is worth his time to study the textbook scene and consult for people across the country.

It is a big, complicated world out there, and there are many useful and exciting things that we can contribute. There comes a point at which your heart may not be in proving a new theorem. You may be tired of the big egos and the competitiveness. Be aware that there are a number of alternatives.

7.10 How to Know When You Are Done for the Day

Reviewing what was said earlier, if you can manage your time well then, at the beginning of each day, you will set realistic goals for yourself. And at the end of the day you will have achieved them. As a result, you will feel that you have done an honest day's work, and you can go home and relax. This might mean taking your spouse or partner out to dinner, or going to a concert, or watching a movie, or reading a book. It could also mean reading some mathematics that is not directly related to anything that is sizzling on your plate right now, or browsing the Web for new ideas. It could mean planning a hiking trip for the coming weekend or setting up lunch with a friend. It does not have to mean hardcore mathematics, and you are no less of a mathematician for not spending every waking hour at the altar making burnt offerings.

If you put in four or five hours a day actually doing intense mathematical research, and another five hours dancing around the fire, then you have done more than it is reasonable to offer to this goddess, and more than will produce good effect. John E. Littlewood always made a point of taking Sunday off. He would take a hike, or spend time with family, or do something to relax his cranium. He always found, then, that on Monday he had a good idea. Perhaps there is a lesson to be learned here. Littlewood wrote more than a hundred papers with Hardy, and his total production over a lifetime was awesome. But he still took Sundays off.

All of this comes down once again to time management. If you use your time well, then you will feel at the end of each day that you have done something worthwhile, that you have actually *accomplished something*. You will not feel as though you have to sweat the night away in some neurotic angst. You can begin the next day fresh and ready for new challenges. That is the way it should be. Being a mathematician does *not* have to be akin to being a tormented *artiste*. You can actually be like everyone else—just a little more inspired and perhaps a little more driven.

7.11 Managing Your Life

I know mathematicians—good ones—who sit in their offices from 8:00 a.m. to 5:00 p.m. each day, go home and have dinner with their families, and then go back to the office for a few (or sometimes several) more hours. I frankly don't understand the point of this habit. There is no evidence that these people are any more productive than anyone else. And surely they have a workspace at the house. Why can't they work at home, surrounded by their families, in the evenings? I would think it would put a strain on any relationship to have a partner who is married to his/her job to such a degree. And I repeat: Where is the evidence that this leads to greater productivity?

You need to strike a balance between your professional life and your personal life. One of the reasons that personal relationships in Hollywood are so unstable is that people are simply unable to differentiate between their public persona and their private persona. It is your responsibility to differentiate between your Gestalt as a mathematician and your Gestalt as a person. You owe something to your profession, but you also owe something to your spouse or partner. And of course to your children. I was talking to a mathematician a while ago about why he got divorced. His answer was, "Early on in the marriage I told my wife my priorities: number one was mathematics; number two was my kids; and she came in number three. My wife accepted this, and we lived happily together for several years. But at some stage I took up running, and my wife could see that she was slipping into fourth place." Well, this was an honest evaluation of his life, and it led to a predictable outcome. If you go back to the office each night after dinner, then perhaps you are sending your spouse or partner a similar signal.

As I have said elsewhere in this book, if you can put in a good piece of time on mathematics each day, then you will be a productive and successful mathematician. No doubt about it. Spending twenty hours a day playing at being a mathematician will not make you any better, and certainly not make you any more successful. A little self-knowledge will take you a long way here. Decide what is important in your life and act on that insight. You will be a happier person as a result.

Glossary

As with any sophisticated professional activity, being a mathematician carries with it a certain amount of verbal baggage. There are all sorts of terms that we commonly bandy about that would mystify the average layman. And, if you are new to this business, they may mystify you as well. So I include most of them here for your delectation. Definitions are the author's own, and the reader may find variants in the literature.

AAAS See *American Association for the Advancement of Science.*

AAUP See *American Association of University Professors.*

A.B.D. Abbreviation for "all but dissertation" (also called "A.B.T.," or "all but thesis"). The phrase describes a student who has completed all parts of the Ph.D. program *except* for the dissertation. At many schools, "A.B.D." is an official status, and you fill out some paperwork to ratify the fact that you have done everything but the dissertation. A great many students leave graduate school at the A.B.D. stage and never complete the degree.

Abel Prize Created in 2002 by the Norwegian government, this prize is modeled after the Swedish Nobel Prize and is intended to recognize excellent work in mathematics.

A.B.T. See *A.B.D.*

academic integrity The rules of conduct by which we live academic life. These include not to cheat on exams and not to plagiarize. Also one should respect the work of others.

Academic Senate A governing body of the university, usually peopled by elected members of the faculty. Also called the *Faculty Senate.*

ACM See *Association for Computing Machinery.*

actuary   A mathematical scientist who calculates annuities, amortization plans, and other insurance data.

adjunct faculty   Teaching faculty, usually those who are hired to teach specific, individual courses. Such people are paid by the course, and usually have no benefits.

Administrative Assistant   See *Head Secretary*.

Allendoerfer Award   Beginning in 1976, this award was created for papers in *Mathematics Magazine*. The prize is administered by the Mathematical Association of America.

American Association for the Advancement of Science (AAAS)   This is an international, nonprofit organization dedicated to advancing science around the world by serving as an educator, leader, spokesperson, and professional association.   In addition to organizing membership activities, AAAS publishes the journal *Science*, as well as many scientific newsletters, books, and reports.

American Association of University Professors (AAUP)   This is a labor organization for university professors—a little bit like a union. It looks after the rights and privileges of those in the professorial ranks. It seeks to protect tenure and academic freedom.

American Mathematical Monthly   A primary mathematics journal of the Mathematical Association of America. See the URL
`http://www.maa.org/pubs/monthly.html`.

American Mathematical Society (AMS)   A professional organization of mathematicians that is primarily interested in research and its attendant activities. Publishes many important books and journals and organizes significant meetings.

American Statistical Association (ASA)   One of several professional mathematical associations in the United States. The purpose of the ASA is to support and promote statistical activities and scholarship. Consult the URL   `http://www.amstat.org`.

AMS Cover Sheet   A standard information sheet, available on the Internet and also in issues of the *Notices of the American Mathematical Society*, to be included with job application materials.

AMS Employment Center   The job interview activities sponsored by the AMS/MAA/SIAM at the January annual meeting.

ASA   See *American Statistical Association*.

ASL See *Association for Symbolic Logic.*

Assistant Professor This is the junior level in the academic ranks. After six years, an Assistant Professor will be considered for tenure and promotion to Associate Professor.

Associate Professor This is the middle level in the academic ranks, between Assistant Professor and Full Professor. An Associate Professor is tenured, and can participate in most departmental decisions (including tenure decisions for junior faculty). An Associate Professor will have more responsibilities than an Assistant Professor.

Association for Computing Machinery (ACM) A national organization that is "a major force in advancing the skills of information technology professionals and students". The Association engages in publishing and organizes conferences. See the URL `http://www.acm.org`.

Association for Symbolic Logic (ASL) A national mathematical organization that concerns itself with fostering and promoting logic and issues that are of concern to logicians. The Association publishes journals and books and organizes conferences. See the URL `http://www.aslonline.org`.

Association for Women in Mathematics (AWM) A professional organization of mathematicians that promotes the interests of women.

Bachelor of Arts Degree Also called the B.A. This is usually a four-year degree, and is the capstone of the undergraduate experience. The degree requires about 120 credit hours of course work and sometimes a thesis. A B.A. degree is usually in the humanities, the arts, or sometimes the social sciences. See *B.A. degree.*

Bachelor of Science Degree Also called the B.S. This is usually a four-year degree, and is the capstone of the undergraduate experience. The degree requires about 120 credit hours of course work and sometimes a thesis. A B.S. degree is usually in the sciences or engineering. See *Bachelor of Science degree.*

B.A. degree The Bachelor of Arts degree. This is usually a four-year degree, and is the capstone of the undergraduate experience. The degree requires about 120 credit hours of course work and sometimes a thesis. A B.A. degree is usually in the humanities, the arts, or sometimes the social sciences. See *Bachelor of Arts degree.*

Bachelor's degree A degree marking the completion of the first four years of *undergraduate* study.

Stefan Bergman Prize   A prize of the American Mathematical Society for excellent research in complex analysis.

George David Birkhoff Prize   A prize to honor excellent work in applied mathematics.

Board of Trustees   This is a collection of businesspeople and prominent social leaders who govern the university or college at a very high level. The Board of Trustees approves all tenure appointments, approves the budget, and engages in other governing activities.

Bôcher Prize   An award given by the American Mathematical Society in recognition of excellent work in analysis.

bottom line   The total or aggregate line from a budget. The phrase is often used to denote the concluding thought from any discussion.

breadth requirement   Courses not in the major field that a student must complete in order to ensure his/her familiarity with a variety of modes of discourse. A key part of a *liberal arts* education.

Chair   Either the *Chairperson* or the *Head*.

Chairperson   A department Chairperson is a leader among equals. The Chairperson is supposed to implement policies that are made by the departmental faculty as a whole. Like a Head, the Chairperson is answerable to the Dean, but he/she also serves the overall faculty. Often a Chairperson is "selected" by a departmental vote or mandate. Nonetheless, the Dean chooses and appoints the Chairperson.

Chair Professor   A distinguished rank among senior faculty is the endowed Chair Professor. This is a special position, usually created with private funds from donors. It entails special perks and privileges, and certainly a high salary.

Chancellor   The Chancellor is the Chief Executive Officer of a college or university. One could also be the Chancellor of an entire university system. Whereas the Provost manages the internal operations of the institution, the Chancellor is in charge of the interface with the public and the government. Especially at a state or public institution, the Chancellor must work with the legislature to ensure that the university has adequate funding and other public support. See *President*.

Chauvenet Prize   A prize, sponsored by the Mathematical Association of America, to recognize excellent mathematical exposition.

College A four-year institution of higher learning that grants Bachelor's degrees and perhaps Master's degrees. Usually a college does not grant the Ph.D. Harvard University has inside it an institution called "Harvard College" which grants undergraduate degrees.

Frank N. Cole Prize An award given by the American Mathematical Society in recognition of excellent work in algebra.

college major At most colleges and universities, an undergraduate will have a field of concentration. Thus the student will take a certain number of courses of general study—just to ensure some background in the humanities, the arts, the social sciences, and the laboratory sciences—before picking a particular subject area in which to concentrate. The student will typically take twenty class hours or more in the major subject area.

college minor In addition to a *college major*, an undergraduate student today will often have a second field of concentration called a "minor". This will be a subject, perhaps related to the major area, in which the student will take several courses and perhaps even write an undergraduate thesis.

colloquium A formal lecture, usually given by a member of another department and often by a professor from another university, that is given for the benefit of the entire math department. The lecture is usually preceded by a ceremonial tea, and there is often a celebratory dinner and even a party afterward.

comprehensive university These institutions started out as the "normal schools", that is, schools that were dedicated to teacher training. Seventy-five years ago there were hundreds of these throughout the country. Today most of these institutions have changed their names, and in some cases, their missions. The primary mission of a comprehensive university today is teaching at the undergraduate and Master's levels.

computer system manager These days most every math department needs somebody to manage the computer system. This includes *e*-mail, the Internet, software installation and maintenance, and hardware installation and maintenance. Ideally, this is a full-time person who is at the service of the department.

Conant Prize A prize, awarded by the American Mathematical Society, to recognize an excellent expository paper in the *Bulletin of the AMS* or the *Notices of the AMS*.

Concerns of Young Mathematicians (CYM) An Internet periodical devoted to issues that are of interest to those beginning in the mathematics profession. See the URL `http://www.youngmath.net/concerns`.

Conference Board of Mathematical Sciences (CBMS)   A national board, a subsidiary of the National Academy of Sciences, that oversees the welfare of the mathematical enterprise in this country. Of particular note are the CBMS conferences. See the Websites
`http://www7.nationalacademies.org/bms/`
and `http://www.cbmsweb.org`.

course work   In the context of a Ph.D. program, this is an explicit requirement that the student take a certain number of credit hours of courses. Often the number is thirty-six hours, but it can be more. Some of these credit hours may be filled with *independent study* courses.

Crafoord Prize   A prize analogous to the Nobel Prize, also awarded by the Swedish academy. It recognizes fields complementary to those honored by the Nobel, including astronomy, mathematics, geosciences, and biosciences.

Curriculum Vitae (CV)   The analogue of what business people call a résumé. This document provides your personal and professional information. It is part of any job application that you may submit.

CV   See *Curriculum Vitae.*

CYM   See *Concerns of Young Mathematicians.*

DARPA   See *Defense Advanced Research Projects Agency.*

Dean   The Dean is a university administrator who sits above all the department Chairs. For example, the Dean of Arts & Sciences at a university runs the Arts & Sciences program. He/she will often manage twenty-five departments. Such a Dean may have an annual budget of $150 million or more.

Defense Advanced Research Projects Agency (DARPA)   A branch of the Central Intelligence Agency that is dedicated to defense-related research. The Agency supports many research projects in mathematics.

Department Administrator   See *Office Manager.*

departmental service   Service by faculty on departmental committees; service by faculty in teaching evaluation; service by faculty in curriculum development; service by faculty in hiring and promotion. And there are many other activities of this nature. Departmental service (and university service) figure into all tenure and promotion decisions.

Department of Energy (DOE) An agency of the U.S. Federal Government that is concerned with energy issues and research in parts of science that impact on energy. In recent years, DOE has been a significant source of funding for mathematical research. See the URL `http://www.eia.doe.gov`.

dissertation Also called the *thesis*. The magnum opus of a Ph.D. program, this document (often seventy-five pages or more) is the student's disquisition on a subject of original research.

Doctorate A *Ph.D.* or other degree at that level. Medicine and Law and the Arts also have Doctoral degrees.

DOE See *Department of Energy*.

Doob Prize A prize, sponsored by the American Mathematical Society, to recognize an excellent mathematics book with wide appeal.

elite private university These are generally privately-funded institutions with large endowments. They receive little or no direct funding from the state or federal government, and are therefore quite independent in their policies and educational procedures.

EIMS See *Employment Information in the Mathematical Sciences*.

Employment Information in the Mathematical Sciences (EIMS) A hard-copy publication of the mathematical societies in which current job openings are advertised. See the URL
`http://www.ams.org/eims/eims-search.html`.

Endowed Chair Professor A Professor for whom the salary, travel funds, and other perks of the position come from a special, endowed fund. It is a great honor to be an Endowed Chair Professor.

Euler Prize Named after Leonhard Euler, this prize was endowed by Paul and Virginia Halmos to recognize an outstanding book about mathematics. The prize was first awarded in 2007.

faculty, disciplining of It happens occasionally that a faculty member will have to be disciplined. This could entail various penalties that must be paid. Usually the Chair, in consultation with the Dean, will handle minor infringements. Major infractions will involve the Provost.

faculty meeting A gathering of the faculty for the purpose of making departmental decisions.

Faculty Senate One of many governing bodies at a college or university. The membership will consist of elected faculty. Also called the *Academic Senate*.

fellowship A grant to pay some or all expenses and support for either a graduate student or a faculty member.

Fields Medal The highest honor in the mathematics profession, awarded every four years to between two and four mathematicians under the age of forty.

final oral See *thesis defense*.

FOCUS The newsletter of the Mathematical Association of America.

Lester R. Ford Award These awards were established in 1964 to recognize authors of articles of expository excellence published in *The American Mathematical Monthly* or *Mathematics Magazine*. The award is administered by the Mathematical Association of America.

freeway flier A person who makes his/her living by having part-time jobs at several different colleges or universities.

Full Professor This is the senior level in the academic ranks. It is the Full Professors who run a department. They decide who gets tenured and who gets promoted to Full Professor. They have a strong voice in hiring and other key decisions.

fungible Funds in a budget are fungible if they can easily be moved from one category to another. For example, funds for foreign travel can often be moved to domestic travel.

generals See *qualifying exams*.

genome project The project, which received massive federal funding, to map the human genome. This project is now substantially completed.

graduate school An educational program that follows upon the usual four-year American undergraduate educational experience. Among other degrees, the graduate program will offer the Master's degree, the Ph.D., or both.

grant This is a quantity of money, provided by a Federal agency or perhaps a private foundation or even by the university, to subsidize a research activity or a curricular project or an educational enterprise. One usually applies for a grant through a formalized procedure, and there is intense competition for grants.

Haimo Award A national award for excellent teaching.

Head A department Head is appointed by the Dean to run a department. Unlike a *Chairperson*, the Head is rather math autonomous. He/she can take advice from the faculty, but will make decisions based on his/her own judgment. Put in other terms, the Head is answerable only to the Dean. See *Chairperson*.

Head Secretary   Also called an Administrative Assistant, an Administrative Aid, or an Administrative Head. This is the chief of all the staff in your department. This person oversees all the secretaries and other staff. Also it is typical for this person to be a budgetary officer and to handle other high-level administrative duties. This person is a direct assistant to the Chair.

Hughes Aircraft   Founded by aviator and entrepreneur Howard Hughes, this is one of the big aircraft manufacturers in Los Angeles. Hughes employs a good many math Ph.D.s from UCLA.

Humboldt Fellowship   A fellowship sponsored by the Humboldt Foundation that promotes cooperation and collaboration between German scholars and non-German scholars.

ICM   See *International Congress of Mathematicians*.

independent study   An arrangement made between a faculty member and a student (either undergraduate or graduate) whereby the student will study a subject area on his/her own and perhaps meet with the faculty member once per week to discuss progress and to ask questions. Usually a student will receive course credit for an independent study.

IMS   See *Institute of Mathematical Statistics*.

IMU   See *International Mathematical Union*.

Institute of Mathematical Statistics   A professional organization located in Beachwood, Ohio. The purpose of the institute is to foster the development and dissemination of the theory and applications of statistics and probability.

instructorship   This is a temporary, usually two- or three-year job for a beginning Ph.D. An instructor spends time at a major research institution in order to be exposed to some important ongoing research programs and some important senior mathematicians. The instructorship will also have some teaching duties. After the instructorship, he/she will apply for an Assistant Professorship (usually at another institution). Also called a *postdoc*.

International Congress of Mathematicians (ICM)   An international gathering of mathematicians, held every four years. This meeting is organized by the International Mathematical Union. It is where the Fields Medals are awarded. It is an important event at which the status and progress of the field are assessed. The ICM was held in Madrid in 2006 and will be held in Hyderabad in 2010.

International Mathematical Union (IMU)   This is the international aggregation of mathematical scholars—in effect the union of all the national mathematical societies. The IMU considers broad issues that affect mathematicians worldwide. It organizes and holds the International Congress of Mathematicians every four years. It awards the Fields Medals. See the URL `http://elib.zib.de/IMU/`.

invited talk at the ICM   An invited 45-minute talk at the International Congress of Mathematicians. A definite distinction for mathematicians. See also *plenary talk at the ICM*.

IT Sector   The Information Technology Sector. This is the computer industry and other allied industries that center around Silicon Valley activities.

journal   A periodical publication in which mathematical research is published. There are also journals that are devoted to teaching and to exposition.

junior college   A college with a two-year curriculum leading to an Associate of Arts (or A.A.) degree. Junior colleges do *not* grant Bachelor's degrees. Junior colleges often act as feeder schools to the large state universities.

large state universities   Ever since the late nineteenth century, states in the U.S. have had a well-developed system of publicly supported universities. In many states these are very large institutions. For families of modest means, the state university is the default place to send their children for higher education.

leave   See *unpaid leave*.

liberal arts college   A four-year institution of higher learning that concentrates on giving students a well-rounded education in the humanities and the arts as well as a particular field of concentration.

Librarian for the department   Ideally the math department should have its own librarian. There are more than 1,700 math journals and many thousands of math books and other resources that need to be at the fingertips of the mathematicians in the department. Especially with all the different types of electronic media, and all the different choices that there are today, it is essential to have a professional in charge of information management.

line item   This is a yearly provision in your departmental budget for a particular standing need. For example, there could be a line item in your departmental budget for $10,000 per year for computer equipment.

M.A. See *Master of Arts degree.*

MAA See *Mathematical Association of America.*

MacArthur Prize A prize to recognize excellence in all fields of human endeavor. This prize is typically five years salary; it is an immensely prestigious and lucrative encomium.

major professor The Professor who directs a Ph.D. thesis. See *thesis advisor.*

Master of Arts Degree A first-level graduate degree, usually in the humanities, the arts, or the social sciences. Earning of this degree may entail passing a course requirement, the taking of some qualifying exams, and possibly writing an expository thesis.

Master of Science Degree A first-level graduate degree, usually in science or engineering. Earning of this degree may entail passing a course requirement, the taking of some qualifying exams, and possibly writing an expository thesis.

Master's Degree A postgraduate degree, usually requiring two to three years of study. This will be in a particular field, like mathematics or chemistry. A Master's degree is intermediate between a Bachelor's degree and a Doctorate. Earning of this degree may entail passing a course requirement, the taking of some qualifying exams, and possibly writing an expository thesis.

maternity leave A leave of absence from work—often with partial or full pay—so that a female parent may assist in the care of a newborn child.

Mathematical Association of America (MAA) A professional organization of mathematicians that is primarily interested in teaching and exposition. Publishes a number of important books and journals and also organizes meetings.

Mathematical Reviews The hard-copy archiving tool of the American Mathematical Society, extant since 1940. This periodical records all papers published in all the major journals, together with complete bibliographic information and a concise review. See the URL
`http://www.ams.org/mr-database`.

`MathSciNet` This is the OnLine version of *Mathematical Reviews*. It is currently a twenty Gigabyte database Internet resource for locating mathematical papers and reviews thereof.

matriculation   The meaning of this word has changed over time, and is also different from country to country. Traditionally matriculation was a ceremony in which the freshman at the university was given a set of exams to gauge qualification for study. When the student passed, he/she was then matriculated. Today it is more common to use the word "matriculation" to mean registration at the university.

media   The physical devices on which we record our ideas for a paper or book or our presentation for a lecture. Traditionally the primary media were different types of paper or cardboard. Today they could be an overhead slide or a disc or tape or cartridge or flash drive or other mass storage device.

mentor   A senior individual who imparts advice, and the wisdom of experience, to a neophyte.

E. H. Moore Prize   A prize to recognize an outstanding research article in one of the AMS primary research journals.

Morgan Prize   A prize to recognize excellent undergraduate research.

M.S.   See *Master of Science Degree*.

National Academy of Sciences (NAS)   A distinguished organization of scientists in the U.S. whose mission is to advise the president in matters of science. More broadly, the NAS helps to set science policy for the country. There are 117 mathematicians in the NAS, and it is a great honor and distinction to be elected to this body.

National Research Council (NRC)   A federal organization that oversees research programs for the government. See the URL `http://www.nas.edu`.

National Research Council (NRC) Group Rankings   The National Research Council's ranking of mathematics Ph.D. programs entails, among other things, a grouping of mathematics Ph.D. programs into Group I, Group II, ..., up to Group V. Departments are ranked according to several characteristics, the main one being the scholarly quality of the faculty. In the 1995 ranking, Group I comprises forty-eight departments with scores (in the report [GMF]) in the range 3.00–5.00. Group II comprises fifty-six departments with scores in the range 2.00–2.99. Group III comprises seventy-two departments with scores in the range 0–1.99. Group IV lists doctoral programs in statistics, biostatistics, and biometrics. Group V lists doctoral programs in applied mathematics and applied science.

National Science Foundation (NSF)   An agency of the federal government that is dedicated to the support of basic research in science. Probably the greatest supporter of research in pure mathematics.

National Security Agency (NSA) A federal agency that works in mathematical areas pertaining to the security of the nation. A particular specialization is cryptography. In addition to being a likely place of employment for Ph.D. mathematicians, the NSA is also a source of research funding. See the URL `http://www.nsa.gov`.

Notices of the American Mathematical Society A primary journal of the American Mathematical Society. This is *not* a research journal. Rather, it is the official organ of the AMS. It contains a great deal of society news. See the URL `http://www.ams.org/notices`.

NRC See *National Research Council.*

NSA See *National Security Agency.*

NSF See *National Science Foundation.*

NSF Postdoctoral Fellowship A fellowship program of the National Science Foundation designed to permit participants to choose research environments that will have maximal impact on their future scientific development. The program provides twenty-four months of support for each awardee. See the URL `http://www.nsf.gov/funding/pgm_summ.jsp?pims_id=5301`.

Office Manager The senior staff person in the department. Sometimes called the *department administrator.*

orals See *qualifying exams.*

Ostrowski Prize A mathematics award given every other year by an international jury from the universities of Basel, Jerusalem, Waterloo, and the academies of Denmark and the Netherlands. The prize, founded in 1989, is funded by the estate of Alexander Ostrowski.

outside offer A job offer from an institution other than your home institution—the one where you have a job right now. An outside offer can be generated because of your research qualifications, or your teaching qualifications, or your ability to have an impact on infrastructure. For a senior faculty member, an outside offer could be generated to attract him/her to be Chair.

paternity leave A leave of absence from work—often with partial or full pay—so that a male parent may assist in the care of a newborn child. Usually this requires the male parent proving that he is providing more than half the care.

Ph.D. The highest academic degree that is granted by most universities in the United States. It is based on *course work, qualifying exams,* and a *thesis* that contains original research.

Ph.D. advisor See *thesis advisor.*

Ph.D. Candidate A student who has passed the qualifying exams and has a thesis advisor and thesis problem.

Ph.D. thesis See *thesis.*

PI See *Principal Investigator.*

plenary talk at the ICM An invited one-hour talk at the International Congress of Mathematicians. Considered to be a great honor.

political correctness The idea that one should use language carefully, so as not to abuse or oppress others who cannot defend themselves. An idea that has taken strong hold in universities.

postdoc Also called a "postdoctoral position". See *instructorship.*

President The Chief Executive Officer of a University or College. See *Chancellor.*

Principal Investigator The primary scientist on a proposal to the National Science Foundation or other granting agency. The person responsible for the work proposed in the grant. Also called the PI.

professional society An organization whose purpose is to sustain and promote the profession (i.e., mathematicians). The American Mathematical Society, the Mathematical Association of America, the Society for Industrial and Applied Mathematics, and the Association for Women in Mathematics are four (but certainly not all) important professional societies in the mathematics profession.

Professor See *Assistant Professor, Associate Professor,* and *Full Professor.*

Professor Emeritus This is a retired Professor who is granted certain privileges—such as use of the staff and the library and the computer system. It is something of an honor to be deemed an Emeritus Professor, as opposed to just a retired Professor.

Project NExT A consortium of young mathematicians across the country which gathers regularly to share concerns and to help bring beginners in the profession up to speed. Project NExT particularly stresses networking among its participants. See the URL `http://archives.math.utk.edu/projnext/`.

Provost The Provost at a university sits above all the Deans. Typically the Provost is referred to as the "chief academic officer" of the university. The Provost runs the day-to-day operations of the institution.

publishing a paper The process by which one writes up a paper with original results, submits it to a journal, undergoes the refereeing process, and then has the work appear in a journal.

"Publish or perish" A byword of American higher education for the past half-century, this phrase encapsulates the sentiment that one must engage in academic research and publishing in order to obtain tenure and to flourish as a professor.

Putnam Exam Sponsored by the Putnam Stock Fund, the William Lowell Putnam exam is a nationally competitive math competition that has run continuously since 1938. The exam is six hours, and has twelve questions. The top prize is a fellowship to Harvard University.

qualifying exams A set of exams that students must take in the major subject areas of mathematics—often these are geometry, algebra, and analysis (although some schools may allow other subjects as well)—in order to qualify to write a Ph.D. thesis. Usually these are written exams, though at some schools they could be oral.

recuse If a member of a committee has a conflict of interest on a particular issue, then he/she will recuse himself/herself from the discussions—meaning that he/she will not take part. The person will also not vote on that decision.

research Scholarly activity that involves developing new ideas or discovering and establishing new truths. Research is a large part of scholarly life.

research university An educational institution that grants undergraduate degrees (the B.A. and/or the B.S.) and also graduate degrees (the M.A., M.S., and Ph.D.). Such an institution will also have a faculty that has a vigorous research program, publishes widely, and visits universities all over the world (to collaborate with their faculties and to lecture about their work).

reward system The value system by means of which faculty are rewarded. This would entail a consideration of teaching, research, and departmental service—weighted in an appropriate fashion.

David P. Robbins Prize A prize to recognize excellent work in algebra, combinatorics, or discrete mathematics.

sabbatical Many institutions of higher education, especially research universities, offer to their faculty the opportunity to take a paid leave of absence every seven years. At some schools this privilege is guaranteed. At others one must compete for a sabbatical.

Satter Prize   A prize to recognize outstanding research by a woman.

seminar   This is a working group that usually meets once a week to learn some focused subject area of mathematics. Most of the time, the seminar speakers will be members of the home math department.

service   Every tenure-track faculty member will have duties in the department, ranging from committee service, to acting as Vice-Chair, to supervising a task force. The aggregate of these activities is referred to as *service*.

sexual harassment   A legal term referring to abuse of power to gain sexual favors. Punishable by prison term or other severe sanctions.

SIAM   See *Society for Industrial and Applied Mathematics*.

Silicon Valley   Centering around Los Gatos, California (where the Apple Computer was invented), this is today one of the hubs of the computer and technology industry.

Sloan Foundation Fellowship   Fellowships given to enhance the careers of the very best young faculty in the sciences. The Sloan Fellowship enables a young mathematician to take a one-year (or longer) leave of absence at another research institution. See the URL   `http://www.sloan.org`. In fact the Sloan Foundation supports a variety of academic activities, both for junior and for senior faculty.

Society for Advancement of Chicanos and Native Americans in Science (SACNAS)   A professional society of mathematicians that promotes the interests of Chicanos and Native Americans.

Society for Industrial and Applied Mathematics (SIAM)   A professional organization of mathematicians that promotes the interests of applied mathematicians and mathematicians who work in industry. The society publishes a number of important books and journals and organizes meetings.

soft money   Money obtained from grant funds. These are not part of the regular university budget, and can disappear when the grant expires.

staff   These are the non-academic members of the math department, including secretaries, computer managers, some counselors, and so forth.

startup fund   A fund provided to a new hire (usually by the Dean, but sometimes by the math department itself) to help with setup, equipment purchases, bringing in collaborators, and the like.

state universities   See *large state universities.*

Leroy P. Steele Prize   Three awards given by the American Mathematical Society in recognition of **(a)** a seminal paper, **(b)** an outstanding career contribution, **(c)** writing (usually a book).

subvene   If your university is paying your expenses for a trip then it is subvening your travel.

Systems Administrator   The person in your department who manages the computer system, all software and hardware.

TA   See *teaching assistant.*

teaching   Teaching students—especially undergraduates—is the way that college and university faculty justify their existence to the public and to the administration. This is the one faculty activity that everyone understands.

teaching assistant (TA)   A graduate student who assists in the teaching of university classes. This is usually done as part of the arrangement to justify the financial support of a graduate student.

teaching evaluation   Traditionally these are paper surveys that are distributed to students in a class near the end of the semester to assess the effectiveness of the teaching. These are collected by a student and turned in to the math department office so that the Professor cannot see them until after the semester is completed. Today teaching evaluations are sometimes done OnLine.

teaching load   The number of courses per semester, or per year, that you are required to teach. Mathematicians frequently describe their teaching load as "3-and-2"—meaning that they teach three courses in one semester and two courses in the other. Sometimes people will describe their teaching load in terms of the number of contact hours.

tenure   This is a form of job security that is special to academic life. A tenured faculty member is *very* difficult to terminate. Typical grounds for termination of tenure are **(a)** moral turpitude, **(b)** fiscal indiscretions, **(c)** academic fraud. Tenure was originally created so that faculty could pursue daring or controversial courses of study, or teach "dangerous" ideas in class.

tenure clock   Every Assistant Professor, by the rules of the American Association of University Professors, must be considered for tenure before seven years of his/her contract expire. Thus every Assistant Professor is on a *tenure clock*. When a person accepts a job as an Assistant Professor, there is some negotiation as to how the tenure clock will be initially set. Sometimes the new Assistant Professor gets credit for time served elsewhere, sometimes not.

Tenure Document   Every college or university has a document called the *Tenure Document* which lays out **(i)** what tenure is, **(ii)** how tenure is bestowed, and **(iii)** how tenure can be forfeited. It is not clear to anyone what the legal status of the *Tenure Document* is, but in the context of the university it defines how things work. All tenure decisions are based on this document. All cases of adjudication of tenure are based on this document. Although it is not widely disseminated, it is readily available to all faculty.

tenure dossier   The collection of materials, including teaching evaluations, outside review letters, information about service, a publication list, and other information that pertains to a tenure case.

tenure, loss of   It is quite difficult to lose tenure, but fiscal malfeasance, sexual misconduct, or failure to teach are some of the reasons that this can come about. There are strict and detailed procedures, outlined in the *Tenure Document*, for adjudicating these matters. Every level of the university is involved in such a decision.

tenure-track   Describes those positions in the department that may lead to tenure.

textbook   The book that is used as a resource in a course. Some courses will have more than one textbook. Some courses will use an OnLine textbook. A textbook will often have supplements, such as a Solutions Manual.

thesis   Also called the *dissertation*. The magnum opus of a Ph.D. program, this document (often seventy-five pages or more) is the student's disquisition on original research.

thesis advisor   The professor who directs the Ph.D. thesis. Often the advisor provides the problem that the student works on, and offers advice and encouragement along the way. The thesis advisor will be a tenure-track faculty member.

thesis committee   The committee, chaired by the thesis advisor and made up of members of the math department plus select members of other departments, that adjudicates the validity of the Ph.D. thesis. The thesis committee presides over the *thesis defense*.

thesis defense   The final ceremonial presentation by the Ph.D. candidate of the thesis results to the Ph.D. committee and a select audience.

thesis problem The question or subject area that you will study in order to develop the materials for your Ph.D. thesis.

TIAA-CREF TIAA stands for "Teachers Insurance and Annuity Association" and CREF stands for "College Retirement Equities Fund". These two programs were created by Andrew Carnegie to ensure that America's teachers were well cared for in their golden years. A good many college and university instructors and professors have retirement packages that depend on TIAA-CREF.

two-body problem The problem of a mathematician and his/her partner who are both professionals, and who both seek employment at a new location.

undergraduate This is a student in the first four years of college, studying for a Bachelor's degree (a B.A. or B.S. is typical).

underrepresented group A societal group—women or African-Americans or Native Americans or one of several others—that has a lower percentage representation in the mathematics profession than the percentage of its representation in the American population. The definition of "underrepresented group" is somewhat political, as Asian-Americans do *not* usually constitute an underrepresented group.

university Traditionally this is an institution of higher learning that has the power to grant Ph.D. degrees.

university service Service by faculty on university-wide committees. This may entail teaching evaluation, curriculum development, evaluating tenure cases, administering discipline, or many other duties. University service (and departmental service as well) figures into all tenure and promotion decisions.

unpaid leave This is a leave of absence from your home university or institution in which you are paid entirely by the host institution (and your home institution provides no funds).

Veblen Prize A prize to recognize excellent work in geometry.

Vice-Chair for Graduate Studies This is an administrative post in the department for managing the graduate program. It is usually occupied by a tenured Professor.

Vice-Chair for Undergraduate Studies This is an administrative post in the department for managing the undergraduate program. It is usually occupied by a tenured Professor.

vita See *Curriculum Vitae*.

Norbert Wiener Prize   A prize to recognize excellent work in applied mathematics.

Wolf Prize   A prize to recognize mathematical research in a variety of fields.

Zentralblatt für Mathematik   The European version of `MathSciNet`, published by Springer-Verlag. Today known as **Zentralblatt MATH**. Their digital-age, OnLine version is available for free on the Web at `http://www.emis.de/ZMATH`. See also `http://www.zblmath.fiz-karlsruhe.de/`.

Bibliography

[**BEC**] D. Bennett and A. Cranell, eds., *Starting Our Careers*, American Mathematical Society, Providence, RI, 1999.

[**BRA**] I. M. Bray, *Effective Fundraising for Nonprofits: Real World Strategies that Work*, NOLO, New York, 2005.

[**BRU**] D. Bruff, Valuing and evaluating teaching in the mathematics faculty hiring process, *Notices of the American Mathematical Society* **54**(2007), 1308–1315.

[**CAL**] C. R. Calleros, Reconciliation of civil rights and civil liberties after R.A.V. v. St. Paul: Antiharassment policies, free speech, multicultural education, and political correctness at Arizona State University, 1992 *Univ. Utah L. Rev.*, 1205.

[**CAS**] B. A. Case, ed., *You're the Professor, What's Next?*, Mathematical Association of America, Washington, D.C., 1994.

[**CON**] J. B. Conway, *On Being a Department Head: A Personal View*, American Mathematical Society, Providence, RI, 1996.

[**DAV**] B. G. Davis, *Tools for Teaching*, Jossey-Bass Publishers, San Francisco, 1993.

[**DAVH**] P. J. Davis and R. Hersh, *The Mathematical Experience*, Mariner Books, New York, 1999.

[**EWI**] J. Ewing, Paul Halmos: In his own words, *Notices of the American Mathematical Society* **54**(2007), 1136–1144.

[**FLA**] J. Flanagan, *Successful Fundraising: A Complete Handbook for Volunteers and Professionals*, McGraw-Hill, New York, 2002.

[**GKM**] E. A. Gavosto, S. G. Krantz, and W. McCallum, *Contemporary Issues in Mathematics Education*, Cambridge University Press, Cambridge, 1999.

[**GTM**] C. E. Glassick, M. T. Taylor, and G. I. Maeroff, *Scholarship Assessed: Evaluation of the Professoriate*, Jossey-Bass Publishers, New York, 1997.

[**GMF**] M. L. Goldberger, B. A. Maher, and P. E. Flattau, *Research-Doctoral Programs in the United States: Continuity and Change*, National Academy Press, Washington, D.C., 1995.

[**HAR**] G. H. Hardy, *A Mathematician's Apology*, Cambridge University Press, Cambridge, 1967.

[**HAT**] M. Harris and R. Taylor, *The Geometry and Cohomology of Some Simple Shimura Varieties*, Princeton University Press, Princeton, NJ, 2001.

[**HIL**] M. A. Hiltzik, *Dealers of Lightning: Xerox PARC and the Dawn of the Computer Age*, Harper Collins, New York, 1999.

[**JON**] S. P. Jones, https://research.microsoft.com/users/simonpj/papers/giving-a-talk/giving-a-talk-slides.pdf.

[**KRA1**] S. G. Krantz, *How to Teach Mathematics*, 2nd ed., American Mathematical Society, Providence, RI, 1999.

[**KRA2**] S. G. Krantz, *A Primer of Mathematical Writing*, American Mathematical Society, Providence, RI, 1997.

[**KRA3**] S. G. Krantz, *A Mathematician's Survival Guide*, American Mathematical Society, Providence, RI, 2004.

[**KRA4**] S. G. Krantz, *Mathematical Publishing, a Guidebook*, American Mathematical Society, Providence, RI, 2005.

[**KRA5**] S. G. Krantz, How to write your first paper, *Notices of the American Mathematical Society* **54**(2007), 1507–1511.

[**KRA6**] S. G. Krantz, *Handbook of Typography for the Mathematical Scientist*, CRC Press, Boca Raton, 2001.

[**KRA7**] S. G. Krantz, *Mathematical Apocrypha*, Mathematical Association of America, Washington, D.C., 2002.

[**KRA8**] S. G. Krantz, *Mathematical Apocrypha Redux*, Mathematical Association of America, Washington, D.C., 2006.

[**LAN**] S. Lang, *Introduction to Arakelov Theory*, Springer, New York, 1988.

[**LIT**] J. E. Littlewood, *Littlewood's Miscellany*, edited by Béla Bollobás, Cambridge University Press, Cambridge, 1986.

[**MCC**] J. McCarthy, How to give a good colloquium, American Mathematical Society, `http://www.ams.org/ams/gcoll.pdf`.

[**MOO**] D. S. Moore, The craft of teaching, *Focus* **15**(1995), 5–8.

[**NAS**] S. Nasar, *A Beautiful Mind*, Faber and Faber, New York, 2002.

[**NAG**] S. Nasar and D. Gruber, Manifold destiny, *The New Yorker*, August 28, 2006, 44–57.

[**REZ**] B. Reznick, *Chalking it Up*, Random House/Birkhäuser, Boston, 1988.

[**RIS**] T. W. Rishel, *Teaching First: A Guide for New Mathematicians*, Mathematical Association of America, Washington, D.C., 2000.

[**ROSG**] A. Rosenberg, et al, *Suggestions on the Teaching of College Mathematics*, Report of the Committee on the Undergraduate Program in Mathematics, Mathematical Association of America, Washington, D.C., 1972.

[**SHA**] M. H. Sharobeam and K. Howard, Teaching demands versus research productivity, *Journal of College Science Teaching* **31**(2002), 436–441.

[**SMA**] S. Smale, Finding a horseshoe on the beaches of Rio, *Mathematical Intelligencer* **20**(1998), 39-44.

[**STE**] Norman E. Steenrod, P. R. Halmos, Menahem Schiffer, and Jean A. Dieudonne, *How to Write Mathematics*, American Mathematical Society, Providence, RI, 1973.

[**STSA**] M. Steuben and D. Sanford, *Twenty Years Before the Blackboard*, Mathematical Association of America, Washington, D.C., 1998.

[**STW**] I. Stewart, *Letters to a Young Mathematician*, Basic Books, New York, 2006.

[**SWJ**] H. Swann and J. Johnson, *E. McSquared's Original, Fantastic, and Highly Edifying Calculus Primer*, W. Kaufman, Los Altos, 1975.

[**THU**] W. Thurston, Mathematical education, *Notices of the American Mathematical Society* **37**(1990), 844–850.

[**TBJ**] R Traylor, W. Bane, and M. Jones, *Creative Teaching: The Heritage of R. L. Moore*, University of Houston, 1972.

[**TUF1**] E. Tufte, *The Visual Display of Quantitative Information*, Graphics Press, Cheshire, Connecticut, 1983.

[**TUF2**] E. Tufte, *Envisioning Information*, Graphics Press, Cheshire, Connecticut, 1990.

[**WEI**] A. Weil, *Basic Number Theory*, 2nd ed., Springer-Verlag, Berlin, 1973.

[**WEN**] T. J. Wenzel, What is an appropriate teaching load for a research-active faculty member at a predominantly undergraduate institution?, *Council on Undergraduate Research Quarterly* **21**(2001), 104–107.

[**WIL**] L. Wilson, *The Academic Man: A Study in the Sociology of a Profession*, Octagon Books, New York, 1942.

[**ZUC**] S. Zucker, Teaching at the university level, *Notices of the American Mathematical Society* **43**(1996), 863–865.

[**ZYG**] A. Zygmund, *Trigonometric Series*, Cambridge University Press, Cambridge, 1968.

Index